Deductive Logic

Deductive Logic

by St. George William Joseph Stock

Copyright © 8/27/2015
Jefferson Publication

ISBN-13: 978-1517089955

Printed in the United States of America

Contents

INTRODUCTION.

§ 1. LOGIC is divided into two branches, namely—

(1) Inductive,

(2) Deductive.

§ 2. The problem of inductive logic is to determine the actual truth or falsity of propositions: the problem of deductive logic is to determine their relative truth or falsity, that is to say, given such and such propositions as true, what others will follow from them.

§ 3. Hence in the natural order of treatment inductive logic precedes deductive, since it is induction which supplies us with the general truths, from which we reason down in our deductive inferences.

§ 4. It is not, however, with logic as a whole that we are here concerned, but only with deductive logic, which may be defined as The Science of the Formal Laws of Thought.

§ 5. In order fully to understand this definition we must know exactly what is meant by 'thought,' by a 'law of thought,' by the term 'formal,' and by 'science.'

§ 6. Thought, as here used, is confined to the faculty of comparison. All thought involves comparison, that is to say, a recognition of likeness or unlikeness.

§ 7. The laws of thought are the conditions of correct thinking. The term 'law,' however, is so ambiguous that it will be well to determine more precisely in what sense it is here used.

§ 8. We talk of the 'laws of the land' and of the 'laws of nature,' and it is evident that we mean very different things by these expressions. By a law in the political sense is meant a command imposed by a superior upon an inferior and sanctioned by a penalty for disobedience. But by the 'laws of nature' are meant merely certain uniformities among natural phenomena; for instance, the 'law of gravitation' means that every particle of matter does invariably attract every other particle of matter in the universe.

§ 9. The word 'law' is transferred by a metaphor from one of these senses to the other. The effect of such a command as that described above is to produce a certain amount of uniformity in the conduct of men, and so, where we observe uniformity in nature, we assume that it is the result of such a command, whereas the only thing really known to us is the fact of uniformity itself.

§ 10. Now in which of these two senses are we using the term 'laws of thought'? The laws of the land, it is plain, are often violated, whereas the laws of nature never can be so [Footnote: There is a sense in which people frequently speak of the laws of nature being violated, as when one says that intemperance or celibacy is a violation of the laws of nature, but here by 'nature' is meant an ideal perfection in the conditions of existence.]. Can the laws of thought be violated in like manner with the laws of the land? Or are they inviolable like the laws of nature?

§ 11. In appearance they can be, and manifestly often are violated-for how else could error be possible? But in reality they can not. No man ever accepts a contradiction when it presents itself to the mind as such: but when reasoning is at all complicated what does really involve a contradiction is not seen to do so; and this sort of error is further assisted by the infinite perplexities of language.

§ 12. The laws of thought then in their ultimate expression are certain uniformities which invariably hold among mental phenomena, and so far they resemble the laws of nature: but in their complex applications they may be violated owing to error, as the laws of the land may be violated by crime.

§ 13. We have now to determine the meaning of the expression 'formal laws of thought.'

§ 14. The distinction between form and matter is one which pervades all nature. We are familiar with it in the case of concrete things. A cup, for instance, with precisely the same form, may be composed of very different matter-gold, silver, pewter, horn or what not?

§ 15. Similarly in every act of thought we may distinguish two things—

(1) the object thought about,

(2) the way in which the mind thinks of it.

The first is called the Matter; the second the Form of Thought.

§ 16. Now Formal, which is another name for Deductive Logic, is concerned only with the way in which the mind thinks, and has nothing to do with the particular objects thought about.

§ 17. Since the form may be the same, whilst the matter is different, we may say that formal logic is concerned with the essential and necessary elements of thought as opposed to such as are accidental and contingent. By 'contingent' is meant what holds true in some cases, but not in others. For instance, in the particular case of equilateral triangles it is true to say, not only that 'all equilateral triangles are equiangular,' but also that 'all equiangular triangles are equilateral.' But the evidence for these two propositions is independent. The one is not a formal consequence of the other. If it were, we should be able to apply the same inference to all matter, and assert generally that if all A is B, all B is A, which it is notorious that we cannot do.

§ 18. It remains now for the full elucidation of our definition to determine what is meant by 'science.'

§ 19. The question has often been discussed whether logic is a science or an art. The answer to it must depend upon the meaning we assign to these terms.

§ 20. Broadly speaking, there is the same difference between Science and Art as there is between knowing and doing.

Science is systematized knowledge;

Art is systematized action.

Science is acquired by study;

Art is acquired by practice.

§ 21. Now logic is manifestly a branch of knowledge, and does not necessarily confer any practical skill. It is only the right use of its rules in thinking which can make men think better. It is therefore, in the broad sense of the terms, wholly a science and not at all an art.

§ 22. But this word 'art,' like most others, is ambiguous, and is often used, not for skill displayed in practice, but for the knowledge necessary thereto. This meaning is better conveyed by the term 'practical science.'

§ 23. Science is either speculative or practical. In the first case we study merely that we may know; in the latter that we may do.

Anatomy is a speculative science;

Surgery is a practical science.

In the first case we study the human frame in order that we may understand its structure; in the second that we may assist its needs. Whether logic is a speculative or a practical science depends entirely upon the way in which it is treated. If we study the laws of thought merely that we may know what they are, we are making it a speculative science; if we study the same laws with a view to deducing rules for the guidance of thought, we are making it a practical science.

§ 24. Hence logic may be declared to be both the science and the art of thinking. It is the art of thinking in the same sense in which grammar is the art of speaking. Grammar is not in itself the right use of words, but a knowledge of it enables men to use words correctly. In the same way a knowledge of logic enables men to think correctly, or at least to avoid incorrect thoughts. As an art logic may be called the navigation of the sea of thought.

6

§ 25. The laws of thought are all reducible to the three following axioms, which are known as The Three Fundamental Laws of Thought.

(1) The Law of Identity—
Whatever is, is;
or, in a more precise form,
Every A is A.

(2) The Law of Contradiction—
Nothing can both be and not be;
Nothing can be A and not A.

(3) The Law of Excluded Middle—
Everything must either be or not be;
Everything is either A or not A.

§ 26. Each of these principles is independent and self-evident.

§ 27. If it were possible for the law of identity to be violated, no violation of the law of contradiction would necessarily ensue: for a thing might then be something else, without being itself at the same time, which latter is what the law of contradiction militates against. Neither would the law of excluded middle be infringed. For, on the supposition, a thing would be something else, whereas all that the law of excluded middle demands is that it should either be itself or not. A would in this case adopt the alternative of being not A.

§ 28. Again, the violation of the law of contradiction does not involve any violation of the law of identity: for a thing might in that case be still itself, so that the law of identity would be observed, even though, owing to the law of contradiction not holding, it were not itself at the same time. Neither would the law of excluded middle be infringed. For a thing would, on the supposition, be both itself and not itself, which is the very reverse of being neither.

§ 29. Lastly, the law of excluded middle might be violated without a violation of the law of contradiction: for we should then have a thing which was neither A nor not A, but not a thing which was both at the same time. Neither would the law of identity be infringed. For we should in this case have a thing which neither was nor was not, so that the conditions of the law of identity could not exist to be broken. That law postulates that whatever is, is: here we have a thing which never was to begin with.

§ 30. These principles are of so simple a character that the discussion of them is apt to be regarded as puerile. Especially is this the case with regard to the law of identity. This principle in fact is one of those things which are more honoured in the breach than in the observance. Suppose for a moment that this law did not hold—then what would become of all our reasoning? Where would be the use of establishing conclusions about things, if they were liable to evade us by a Protean change of identity?

§ 31. The remaining two laws supplement each other in the following way. The law of contradiction enables us to affirm of two exhaustive and mutually exclusive alternatives, that it is impossible for both to be true; the law of excluded middle entitles us to add, that it is equally impossible for both to be false. Or, to put the same thing in a different form, the law of contradiction lays down that one of two such alternatives must be false; the law of excluded middle adds that one must be true.

§32. There are three processes of thought

(1) Conception.

(2) Judgement.

(3) Inference or Reasoning.

§ 33. Conception, which is otherwise known as Simple Apprehension, is the act of forming in the mind the idea of anything, e.g. when we form in the mind the idea of a cup, we are performing the process of conception.

§ 34. Judgement, in the sense in which it is here used [Footnote: Sometimes the term 'judgement' is extended to the comparison of nameless sense-impressions, which underlies the formation of concepts. But this amounts to identifying judgement with thought in general.] may be resolved into putting two ideas together in the mind, and pronouncing as to their agreement or disagreement, e.g. we have in our minds the idea of a cup and the idea of a thing made of porcelain, and we combine them in the judgement—'This cup is made of porcelain.'

§ 35. Inference, or Reasoning, is the passage of the mind from one or more judgements to another, e.g. from the two judgements 'Whatever is made of porcelain is brittle,' and 'This cup is made of porcelain,' we elicit a third judgement, 'This cup is brittle.'

§ 36. Corresponding to these three processes there are three products of thought, viz.
(1) The Concept.
(2) The Judgement.
(3) The Inference.

§ 37. Since our language has a tendency to confuse the distinction between processes and products, [Footnote: E.g. We have to speak quite indiscriminately of Sensation, Imagination, Reflexion, Sight, Thought, Division, Definition, and so on, whether we mean in any case a process or a product.] it is the more necessary to keep them distinct in thought. Strictly we ought to speak of conceiving, judging and inferring on the one hand, and, on the other, of the concept, the judgement and the inference.

The direct object of logic is the study of the products rather than of the processes of thought. But, at the same time, in studying the products we are studying the processes in the only way in which it is possible to do so. For the human mind cannot be both actor and spectator at once; we must wait until a thought is formed in our minds before we can examine it. Thought must be already dead in order to be dissected: there is no vivisection of consciousness. Thus we can never know more of the processes of thought than what is revealed to us in their products.

§ 38. When the three products of thought are expressed in language, they are called respectively
(1) The Term.
(2) The Proposition.
(3) The Inference.

§ 39. Such is the ambiguity of language that we have already used the term 'inference' in three different senses—first, for the act or process of inferring; secondly, for the result of that act as it exists in the mind; and, thirdly, for the same thing as expressed in language. Later on we shall have to notice a further ambiguity in its use.

§ 40. It has been declared that thought in general is the faculty of comparison, and we have now seen that there are three products of thought. It follows that each of these products of thought must be the result of a comparison of some kind or other.

The concept is the result of comparing attributes.

The judgement is the result of comparing concepts.

The inference is the result of comparing judgements.

§ 41. In what follows we shall, for convenience, adopt the phraseology which regards the products of thought as clothed in language in preference to that which regards the same products as they exist in the mind of the individual. For although the object of logic is to examine thought pure and simple, it is obviously impossible to discuss it except as clothed in language. Accordingly the three statements above made may be expressed as follows—

The term is the result of comparing attributes.

The proposition is the result of comparing terms.

The inference is the result of comparing propositions.

§ 42. There is an advantage attending the change of language in the fact that the word 'concept' is not an adequate expression for the first of the three products of thought, whereas the word 'term' is. By a concept is meant a general notion, or the idea of a class, which corresponds only to a common term. Now not only are common terms the results of comparison, but singular terms, or the names of individuals, are so too.

§ 43. The earliest result of thought is the recognition of an individual object as such, that is to say as distinguished and marked off from the mass of its surroundings. No doubt the first impression produced Upon the nascent intelligence of an infant is that of a confused whole. It requires much exercise of thought to distinguish this whole into its parts. The completeness of the recognition of an individual object is announced by attaching a name to it. Hence even an individual name, or singular term, implies thought or comparison. Before the *child* can attach a meaning to the word '*mother*,' which to it is a singular term, it must have distinguished between the set of impressions produced in it by one object from those which are produced in it by others. Thus, when Vergil says

Incipe, parve puer, risu cognoscere matrem,

he is exhorting the beatific infant to the exercise of the faculty of comparison.

§ 44. That a common term implies comparison does not need to be insisted upon. It is because things resemble each other in certain of their attributes that we call them by a common name, and this resemblance could not be ascertained except by comparison, at some time and by some one. Thus a common term, or concept, is the compressed result of an indefinite number of comparisons, which lie wrapped up in it like so many fossils, witnessing to prior ages of thought.

§ 45. In the next product of thought, namely, the proposition, we have the result of a single act of comparison between two terms; and this is why the proposition is called the unit of thought, as being the simplest and most direct result of comparison.

§ 46. In the third product of thought, namely, the inference, we have a comparison of propositions either directly or by means of a third. This will be explained later on. For the present we return to the first product of thought.

§ 47. The nature of singular terms has not given rise to much dispute; but the nature of common terms has been the great battle-ground of logicians. What corresponds to a singular term is easy to determine, for the thing of which it is a name is there to point to: but the meaning of a common term, like 'man' or 'horse,' is not so obvious as people are apt to think on first hearing of the question.

§ 48. A common term or class-name was known to mediæval logicians under the title of a Universal; and it was on the question 'What is a Universal 7' that they split into the three schools of Realists, Nominalists, and Conceptualists. Here are the answers of the three schools to this question in their most exaggerated form—

§ 49. Universals, said the Realists, are substances having an independent existence in nature.

§ 50. Universals, said the Nominalists, are a mere matter of words, the members of what we call a class having nothing in common but the name.

§ 51. Universals, said the Conceptualists, exist in the mind alone, They are the conceptions under which the mind regards external objects.

§ 52. The origin of pure Realism is due to Plato and his doctrine of 'ideas'; for Idealism, in this sense, is not opposed to Realism, but identical with it. Plato seems to have imagined that, as there was a really existing thing corresponding to a singular term, such as Socrates, so there must be a really existing thing corresponding to the common term 'man.' But when once the existence of these general objects is admitted, they swamp all other existences. For individual men are fleeting and transitory—subject to growth, decay and death—whereas the idea of man is imperishable and eternal. It is only by partaking in the nature of these ideas that individual objects exist at all.

§ 53. Pure Nominalism was the swing of the pendulum of thought to the very opposite extreme; while Conceptualism was an attempt to hit the happy mean between the two.

§ 54. Roughly it may be said that the Realists sought for the answer to the question 'What is a Universal?' in the matter of thought, the Conceptualists in the form, and the Nominalists in the expression.

§ 55. A full answer to the question 'What is a Universal?' will bring in something of the three views above given, while avoiding the exaggeration of each. A Universal is a number of things that are called by the same name; but they would not be called by the same name unless they fell under the same conception in the mind; nor would they fall under the same conception in the mind unless there actually existed similar attributes in the several members of a class, causing us to regard them under the same conception and to give them the same name. Universals therefore do exist in nature, and not merely in the mind of man: but their existence is dependent upon individual objects, instead of individual objects depending for their existence upon them. Aristotle saw this very clearly, and marked the distinction between the objects corresponding to the singular and to the common term by calling the former Primary and the latter Secondary Existences. Rosinante and Excalibur are primary, but 'horse' and 'sword' secondary existences.

§ 56. We have seen that the three products of thought are each one stage in advance of the other, the inference being built upon the proposition, as the proposition is built upon the term. Logic therefore naturally divides itself into three parts.

The First Part of Logic deals with the Term;
The Second Part deals with the Proposition;
The Third Part deals with the Inference.

PART I.—OF TERMS.

CHAPTER 1.

Of the Term as distinguished from other words.

§ 57. The word 'term' means a boundary.

§ 58. The subject and predicate are the two terms, or boundaries, of a proposition. In a proposition we start from a subject and end in a predicate (§§ 182-4), there being nothing intermediate between the two except the act of pronouncing as to their agreement or disagreement, which is registered externally under the sign of the copula. Thus the subject is the 'terminus a quo,' and the predicate is the 'terminus ad quem.'

§ 59. Hence it appears that the term by its very name indicates that it is arrived at by an analysis of the proposition. It is the judgement or proposition that is the true unit of thought and speech. The proposition as a whole is prior in conception to the terms which are its parts: but the parts must come before the whole in the synthetic order of treatment.

§ 60. A term is the same thing as a name or noun.

§ 61. A name is a word, or collection of words, which serves as a mark to recall or transmit the idea of a thing, either in itself or through some of its attributes.

§ 62. Nouns, or names, are either Substantive or Adjective.

A Noun Substantive is the name of a thing in itself, that is to say, without reference to any special attribute.

§ 63. A Noun Adjective is a name which we are entitled to add to a thing, when we know it to possess a given attribute.

§ 64. The Verb, as such, is not recognised by logic, but is resolved into predicate and copula, that is to say, into a noun which is affirmed or denied of another, plus the sign of that affirmation or denial. 'The kettle boils' is logically equivalent to 'The kettle is boiling,' though it is by no means necessary to express the proposition in the latter shape. Here we see that 'boils' is equivalent to the noun 'boiling' together with the copula 'is,' which declares its agreement with the noun 'kettle.' 'Boiling' here is a noun adjective, which we are entitled to add to 'kettle,' in virtue of certain knowledge which we have about the latter. Being a verbal noun, it is called in grammar a participle, rather than a mere adjective. The word 'attributive' in logic embraces both the adjective and participle of grammar.

§ 65. In grammar every noun is a separate word: but to logic, which is concerned with the thought rather than with the expression, it is indifferent whether a noun, or term, consists of one word or many. The latter are known as 'many-worded names.' In the following passage, taken at random from Butler's Analogy—'These several observations, concerning the active principle of virtue and obedience to God's commands, are applicable to passive submission or resignation to his will'—we find the subject consisting of fourteen words, and the predicate of nine. It is the exception rather than the rule to find a predicate which consists of a single word. Many-worded names in English often consist of clauses introduced by the conjunction 'that,' as 'That letters should be written in strict conformity with nature is true': often also of a grammatical subject with one or more dependent clauses attached to it, as

'He who fights and runs away,
Will live to fight another day.'

§ 66. Every term then is not a word, since a term may consist of a collection of words. Neither is every word a term. 'Over,' for instance, and 'swiftly,' and, generally, what are called particles in grammar, do not by themselves constitute terms, though they may be employed along with other words to make up a term.

§ 67. The notions with which thought deals involve many subtle relations and require many nice modifications. Language has instruments, more or less perfect, whereby such relations and modifications may be expressed. But these subsidiary aids to expression do not form a notion which can either have something asserted of it or be asserted itself of something else.

§ 68. Hence words are divided into three classes—

(1) Categorematic;

(2) Syncategorematic;

(3) Acategorematic.

§ 69. A Categorematic word is one which can be used by itself as a term.

§ 70. A Syncategorematic word is one which can help to form a term.

§ 71. An Acategorematic word is one which can neither form, nor help to form, a term [Footnote: Comparatively few of the parts of speech are categorematic. Nouns, whether substantive or adjective, including of course pronouns and participles, are so, but only in their nominative cases, except when an oblique case is so used as to be equivalent to an attributive. Verbs also are categorematic, but only in three of their moods, the Indicative, the Infinitive, and the Potential. The Imperative and Optative moods clearly do not convey assertions at all, while the Subjunctive can only figure as a subordinate member of some assertion. We may notice, too, that the relative pronoun, unlike the rest, is necessarily syncategorematic, for the same reason as the subjunctive mood. Of the remaining parts of speech the article, adverb, preposition, and conjunction can never be anything but syncategorematic, while the interjection is acategorematic, like the vocative case of nouns and the imperative and optative moods of verbs, which do not enter at all into the form of sentence known as the proposition.].

§ 72. Categorematic literally means 'predicable.' 'Horse,' 'swift,' 'galloping' are categorematic. Thus we can say, 'The horse is swift,' or 'The horse is galloping.' Each of these words forms a term by itself, but 'over' and 'swiftly' can only help to form a term, as in the proposition, 'The horse is galloping swiftly over the plain.'

§ 73. A term then may be said to be a categorematic word or collection of words, that is to say, one which can be used by itself as a predicate.

§ 74. To entitle a word or collection of words to be called a term, it is not necessary that it should be capable of standing by itself as a subject. Many terms which can be used as predicates are incapable of being used as subjects: but every term which can be used as a subject (with the doubtful exception of proper names) can be used also as a predicate. The attributives 'swift' and 'galloping' are terms, quite as much as the subject 'horse,' but they cannot themselves be used as subjects.

§ 75. When an attributive appears to be used as a subject, it is owing to a grammatical ellipse. Thus in Latin we say 'Boni sapientes sunt,' and in English 'The good are wise,' because it is sufficiently declared by the inflexional form in the one case, and by the usage of the language in the other, that men are signified. It is an accident of language how far adjectives can be used as subjects. They cease to be logical attributives the moment they are so used.

§ 76. There is a sense in which every word may become categorematic, namely, when it is used simply as a word, to the neglect of its proper meaning. Thus we can say—
'"Swiftly" is an adverb.' 'Swiftly' in this sense is really no more than the proper name for a particular word. This sense is technically known as the 'suppositio materialis' of a word.

CHAPTER II.

Of the Division of Things.

§ 77. Before entering on the divisions of terms it is necessary to advert for a moment to a division of the things whereof they are names.

§ 78. By a 'thing' is meant simply an object of thought—whatever one can think about.

§ 79. Things are either Substances or Attributes. Attributes may be sub-divided into Qualities and Relations.

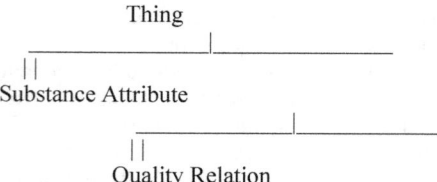

Thing

Substance Attribute

Quality Relation

§ 80. A Substance is a thing which can be conceived to exist by itself. All bodies are material substances. The soul, as a thinking subject, is an immaterial substance.

§ 81. An Attribute is a thing which depends for its existence upon a substance, e.g. greenness, hardness, weight, which cannot be conceived to exist apart from green, hard, and heavy substances.

§ 82. A Quality is an attribute which does not require more than one substance for its existence. The attributes just mentioned are qualities. There might be greenness, hardness, and weight, if there were only one green, hard and heavy substance in the universe.

§ 83. A Relation is an attribute which requires two or more substances for its existence, e.g. nearness, fatherhood, introduction.

§ 84. When we say that a substance can be conceived to exist by itself, what is meant is that it can be conceived to exist independently of other substances. We do not mean that substances can be conceived to exist independently of attributes, nor yet out of relation to a mind perceiving them. Substances, so far as we can know them, are only collections of attributes. When therefore we say that substances can be conceived to exist by themselves, whereas attributes are dependent for their existence upon substances, the real meaning of the assertion reduces itself to this, that it is only certain collections of attributes which can be conceived to exist independently; whereas single attributes depend for their existence upon others. The colour, smoothness or solidity of a table cannot be conceived apart from the extension, whereas the whole cluster of attributes which constitutes the table can be conceived to exist altogether independently of other 'such clusters. We can imagine a table to exist, if the whole material universe were annihilated, and but one mind left to perceive it. Apart from mind, however, we cannot imagine it: since what we call the attributes of a material substance are no more than the various modes in which we find our minds affected.

§ 85. The above division of things belongs rather to the domain of metaphysics than of logic: but it is the indispensable basis of the division of terms, to which we now proceed.

CHAPTER III.

Of the Division of Terms.

§ 86. The following scheme presents to the eye the chief divisions of terms.
 Term
Division of terms according to their place in thought.
 Subject-Term
 Attributive
 according to the kind of thing signified.
 Abstract
 Concrete
 according to Quantity in Extension.
 Singular
 Common
 according to Quality.
 Positive
 Privative
 Negative
 according to number of meanings.
 Univocal
 Equivocal
 according to number of things involved in the name.
 Absolute
 Relative
 according to number of quantities.
 Connotative
 Non-connotative

Subject-term and Attributive.

§ 87. By a Subject-term is meant any term which is capable of standing by itself as a subject, e.g. 'ribbon,' 'horse.'

§ 88. Attributives can only be used as predicates, not as subjects, e.g. 'cherry-coloured,' 'galloping.' These can only be used in conjunction with other words (syncategorematically) to make up a subject. Thus we can say 'A cherry-coloured ribbon is becoming,' or 'A galloping horse is dangerous.'

§ 89. Attributives are contrivances of language whereby we indicate that a subject has a certain attribute. Thus, when we say 'This paper is white,' we indicate that the subject 'paper' possesses the attribute whiteness. Logic, however, also recognises as attributives terms which signify the non-possession of attributes. 'Not-white' is an attributive equally with 'white.'

§ 90. An Attributive then may be defined as a term which signifies the possession, or non-possession, of an attribute by a subject.

§ 91. It must be carefully noticed that attributives are not names of attributes, but names of the things which possess the attributes, in virtue of our knowledge that they possess them. Thus 'white' is the name of all the things which possess the attribute whiteness, and 'virtuous' is a name; not of the abstract quality, virtue, itself, but of the men and actions which possess it. It is clear that a term can only properly be said to be a name of those things whereof it can be predicated. Now, we cannot intelligibly predicate an attributive of the abstract quality, or qualities, the possession of which it implies. We cannot, for instance, predicate the term 'learned' of the abstract quality of learning: but we may predicate it of the individuals, Varro and Vergil. Attributives, then, are to be regarded as names, not of the attributes which they imply, but of the things in which those attributes are found.

§ 92. Attributives, however, are names of things in a less direct way than that in which subject-terms may be the names of the same things. Attributives are names of things only in predication, whereas subject-terms are names of things in or out of predication. The terms 'horse' and 'Bucephalus' are names of certain things, in this case animals, whether we make any statement about them or not: but the terms 'swift' and 'fiery' only become names of the same things in virtue of being predicable of them. When we say 'Horses are swift' or 'Bucephalus was fiery,' the terms 'swift' and 'fiery' become names respectively of the same things as 'horse' and 'Bucephalus.' This function of attributives as names in a secondary sense is exactly expressed by the grammatical term 'noun adjective.' An attributive is not directly the name of anything. It is a name added on in virtue of the possession by a given thing of a certain attribute, or, in some cases, the non-possession.

§ 93. Although attributives cannot be used as subjects, there is nothing to prevent a subject-term from being used as a predicate, and so assuming for the time being the functions of an attributive. When we say 'Socrates was a man,' we convey to the mind the idea of the same attributes which are implied by the attributive 'human.' But those terms only are called attributives which can never be used except as predicates.

§ 94. This division into Subject-terms and Attributives may be regarded as a division of terms according to their place in thought. Attributives, as we have seen, are essentially predicates, and can only be thought of in relation to the subject, whereas the subject is thought of for its own sake.

Abstract and Concrete Terms.

§ 95. An Abstract Term is the name of an attribute, e.g. whiteness [Footnote: Since things cannot be spoken of except by their names, there is a constantly recurring source of confusion between the thing itself and the name of it. Take for instance 'whiteness.' The attribute whiteness is a thing, the word 'whiteness' is a term.], multiplication, act, purpose, explosion.

§ 96. A Concrete Term is the name of a substance, e.g. a man, this chair, the soul, God.

§ 97. Abstract terms are so called as being arrived at by a process of Abstraction. What is meant by Abstraction will be clear from a single instance. The mind, in contemplating a number of substances, may draw off, or abstract, its attention from all their other characteristics, and fix it only on some point, or points, which they have in common. Thus, in contemplating a number of three-cornered objects, we may draw away our attention from all their other qualities, and fix it exclusively upon their three-corneredness, thus constituting the abstract notion of 'triangle.' Abstraction may be performed equally well in the case of a single object: but the mind would not originally have known on what points to fix its attention except by a comparison of individuals.

§ 98. Abstraction too may be performed upon attributes as well as substances. Thus, having by abstraction already arrived at the notion of triangle, square, and so on, we may fix our attention upon what these have in common, and so rise to the higher abstraction of 'figure.' As thought becomes more complex, we may have abstraction on abstraction and attributes of attributes. But, however many steps may intervene, attributes may always be traced back to substances at last. For attributes of attributes can mean at bottom nothing but the co-existence of attributes in, or in connection with, the same substances.

§ 99. We have said that abstract terms are so called, as being arrived at by abstraction: but it must not be inferred from this statement that all terms which are arrived at by abstraction are abstract. If this were so, all names would be abstract except proper names of individual substances. All common terms, including attributives, are arrived at by abstraction, but they are not therefore abstract terms. Those terms only are called abstract, which cannot be applied to substances at all. The terms 'man' and 'human' are names of the same substance of which Socrates is a name. Humanity is a name only of certain attributes of that substance, namely those which are shared by others. All names of concrete things then are concrete, whether they denote them individually or according to classes, and whether directly and in themselves, or indirectly, as possessing some given attribute.

§ 100. By a 'concrete thing' is meant an individual Substance conceived of with all its attributes about it. The term is not confined to material substances. A spirit conceived of under personal attributes is as concrete as plum-pudding.

§ 101. Since things are divided exhaustively into substances and attributes, it follows that any term which is not the name of a thing capable of being conceived to exist by itself, must be an abstract term. Individual substances can alone be conceived to exist by themselves: all their qualities, actions, passions, and inter-relations, all their states, and all events with regard to them, presuppose the existence of these individual substances. All names therefore of such things as those just enumerated are abstract terms. The term 'action,' for instance, is an abstract term. For how could there be action without an agent? The term 'act' also is equally abstract for the same reason. The difference between 'action' and 'act' is not the difference between abstract and concrete, but the difference between the name of a process and the name of the corresponding product. Unless acts can be conceived to exist without agents they are as abstract as the action from which they result.

§ 102. Since every term must be either abstract or concrete, it may be asked—Are attributives abstract or concrete? The answer of course depends upon whether they are names of substances or names of attributes. But attributives, it must be remembered, are never directly names of anything, in the way that subject-terms are; they are only names of things in virtue of being predicated of them. Whether an attributive is abstract or concrete, depends on the nature of the subject of which it is asserted or denied. When we say 'This man is noble,' the term 'noble' is concrete, as being the name of a substance: but when we say 'This act is noble,' the term 'noble' is abstract, as being the name of an attribute.

§ 103. The division of terms into Abstract and Concrete is based upon the kind of thing signified. It involves no reference to actual existence. There are imaginary as well as real substances. Logically a centaur is as much a substance as a horse.

Terms.

§ 104. A Singular Term is a name which can be applied, in the same sense, to one thing only, e.g. 'John,' 'Paris,' 'the capital of France,' 'this pen.'

§ 105. A Common Term is a name which can be applied, in the same sense, to a class of things, e.g. 'man,' 'metropolis,' 'pen.'

In order that a term may be applied in the same sense to a number of things, it is evident that it must indicate attributes which are common to all of them. The term 'John' is applicable to a number of things, but not in the same sense, as it does not indicate attributes.

§ 106. Common terms are formed, as we have seen already (§ 99), by abstraction, i. e. by withdrawing the attention from the attributes in which individuals differ, and concentrating it upon those which they have in common.

§ 107. A class need not necessarily consist of more than two things. If the sun and moon were the only heavenly bodies in the universe, the word 'heavenly body' would still be a common term, as indicating the attributes which are possessed alike by each.

§ 108. This being so, it follows that the division of terms into singular and common is as exhaustive as the preceding ones, since a singular term is the name of one thing and a common term of more than one. It is indifferent whether the thing in question be a substance or an attribute; nor does it matter how complex it may be, so long as it is regarded by the mind as one.

§ 109. Since every term must thus be either singular or common, the members of the preceding divisions must find their place under one or both heads of this one. Subject-terms may plainly fall under either head of singular or common: but attributives are essentially common terms. Such names as 'green,' 'gentle,' 'incongruous' are applicable, strictly in the same sense, to all the things which possess the attributes which they imply.

§ 110. Are abstract terms then, it may be asked, singular or common? To this question we reply—That depends upon how they are used. The term 'virtue,' for instance, in one sense, namely, as signifying moral excellence in general, without distinction of kind, is strictly a singular term, as being the name of one attribute: but as applied to different varieties of moral excellence—justice, generosity, gentleness and so on—it is a common term, as being a name which is applicable, in the same sense, to a class of attributes. Similarly the term 'colour,' in a certain sense, signifies one unvarying attribute possessed by bodies, namely, the power of affecting the eye, and in this sense it is a singular term: but as applied to the various ways in which the eye may be affected, it is evidently a common term, being equally applicable to red, blue, green, and every other colour. As soon as we begin to abstract from attributes, the higher notion becomes a common term in reference to the lower. By a 'higher notion' is meant one which is formed by a further process of abstraction. The terms 'red,' 'blue,' 'green,' etc., are arrived at by abstraction from physical objects; 'colour' is arrived at by abstraction from them, and contains nothing, but what is common to all. It therefore applies in the same sense to each, and is a common term in relation to them.

§ 111. A practical test as to whether an abstract term, in any given case, is being used as a singular or common term, is to try whether the indefinite article or the sign of the plural can be attached to it. The term 'number,' as the name of a single attribute of things, admits of neither of these adjuncts: but to talk of 'a number' or 'the numbers, two, three, four,' etc., at once marks it as a common term. Similarly the term 'unity' denotes a single attribute, admitting of no shades of distinction: but when a writer begins to speak of 'the unities' he is evidently using the word for a class of things of some kind or other, namely, certain dramatical proprieties of composition.

Proper *Names* and *Designations*.

§ 112. Singular terms may be subdivided into Proper Names and Designations.

§ 113. A Proper Name is a permanent singular term applicable to a thing in itself; a Designation is a singular term devised for the occasion, or applicable to a thing only in so far as it possesses some attribute.

§ 114. 'Homer' is a proper name; 'this man,' 'the author of the Iliad' are designations.

§ 115. The number of things, it is clear, is infinite. For, granting that the physical universe consists of a definite number of atoms—neither one more nor one less—still we are far from having exhausted the possible number of things. All the manifold material objects, which are made up by the various combinations of these atoms, constitute separate objects of thought, or things, and the mind has further an indefinite power of conjoining and dividing these objects, so as to furnish itself with materials of thought, and also of fixing its attention by abstraction upon attributes, so as to regard them as things, apart from the substances to which they belong.

§ 116. This being so, it is only a very small number of things, which are constantly obtruding themselves upon the mind, that have singular terms permanently set apart to denote them. Human beings, some domestic animals, and divisions of time and place, have proper names assigned to them in most languages, e.g. 'John,' 'Mary,' 'Grip,' 'January,' 'Easter,' 'Belgium,' 'Brussels,' 'the Thames,' 'Ben-Nevis.' Besides these, all abstract terms, when used without reference to lower notions, are of the nature of proper names, being permanently set apart to denote certain special attributes, e.g. 'benevolence,' 'veracity,' 'imagination,' 'indigestibility, 'retrenchment.'

§ 117. But the needs of language often require a singular term to denote some thing which has not had a proper name assigned to it. This is effected by taking a common term, and so limiting it as to make it applicable, under the given circumstances, to one thing only. Such a limitation may be effected in English by prefixing a demonstrative or the definite article, or by appending a description, e.g. 'this pen,' 'the sofa,' 'the last rose of summer.' When a proper name is unknown, or for some reason, unavailable, recourse may be had to a designation, e.g. 'the honourable member who spoke last but one.'

Collective Terms.

§ 118. The division of terms into singular and common being, like those which have preceded it, fundamental and exhaustive, there is evidently no room in it for a third class of Collective Terms. Nor is there any distinct class of terms to which that name can be given. The same term may be used collectively or distributively in different relations. Thus the term 'library,' when used of the books which compose a library, is collective; when used of various collections of books, as the Bodleian, Queen's library, and so on, it is distributive, which, in this case, is the same thing as being a common term.

§ 119, The distinction between the collective and distributive use of a term is of importance, because the confusion of the two is a favourite source of fallacy. When it is said 'The plays of Shakspeare cannot be read in a day,' the proposition meets with a very different measure of acceptance according as its subject is understood collectively or distributively. The word 'all' is perfectly ambiguous in this respect. It may mean all together or each separately—two senses which are distinguished in Latin by 'totus' or 'cunctus,' for the collective, and 'omnis' for the distributive use.

§ 120. What is usually meant however when people speak of a collective term is a particular kind of singular term.

§ 121. From this point of view singular terms may be subdivided into Individual and Collective, by an Individual Term being meant the name of one object, by a Collective Term the name of several considered as one. 'This key' is an individual term; 'my bunch of keys' is a collective term.

§ 122. A collective term is quite as much the name of one thing as an individual term is, though the thing in question happens to be a group. A group is one thing, if we choose to think of it as one. For the mind, as we have already seen, has an unlimited power of forming its own things, or objects of thought. Thus a particular peak in a mountain chain is as much one thing as the chain itself, though, physically speaking, it is inseparable from it, just as the chain itself is inseparable from the earth's surface. In the same way a necklace is as much one thing as the individual beads which compose it.

§ 123. We have just seen that a collective term is the name of a group regarded as one thing: but every term which is the name of such a group is not necessarily a collective term. 'London,' for instance, is the name of a group of objects considered as one thing. But 'London' is not a collective term, whereas 'flock,' 'regiment,' and 'senate' are. Wherein then lies the difference? It lies in this—that flock, regiment and senate are groups composed of objects which are, to a certain extent, similar, whereas London is a group made up of the most dissimilar objects—streets and squares and squalid slums, fine carriages and dirty faces, and so on. In the case of a true collective term all the members of the group will come under some one common name. Thus all the members of the group, flock of sheep, come under the common name 'sheep,' all the members of the group 'regiment' under the common name, 'soldier,' and so on.

§ 124. The subdivision of singular terms into individual and collective need not be confined to the names of concrete things. An abstract term like 'scarlet,' which is the name of one definite attribute, may be reckoned 'individual,' while a term like 'human nature,' which is the name of a whole group of attributes, would more fitly be regarded as collective.

§ 126. The main division of terms, which we have been discussing, into singular and collective, is based upon their Quantity in Extension. This phrase will be explained presently.

§ 126. We come now to a threefold division of terms into Positive, Privative and Negative. It is based upon an implied two-fold division into positive and non-positive, the latter member being subdivided into Privative and Negative.

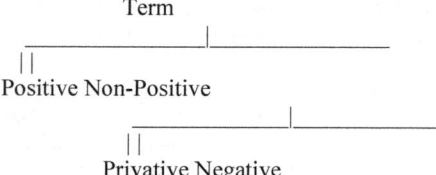

If this division be extended, as it sometimes is, to terms in general, a positive term must be taken to mean only the definite, or comparatively definite, member of an exhaustive division in accordance with the law of excluded middle (§ 25). Thus 'Socrates' and 'man' are positive, as opposed to 'not-Socrates' and 'not-man.'

§ 127. The chief value of the division, however, and especially of the distinction drawn between privative and negative terms, is in relation to attributives.

From this point of view we may define the three classes of terms as follows:

A Positive Term signifies the presence of an attribute, e.g.: 'wise,' 'full.'

A Negative Term signifies merely the absence of an attribute, e.g. 'not-wise,' 'not-full.'

A Privative Term signifies the absence of an attribute in a subject capable of possessing it, e.g. 'unwise,' 'empty'. [Footnote: A privative term is usually defined to mean one which signifies the absence of an attribute where it was once possessed, or might have been expected to be present, e.g. 'blind.' The utility of the slight extension of meaning here assigned to the expression will, it is hoped, prove its justification.]

§ 128. Thus a privative term stands midway in meaning between the other two, being partly positive and partly negative—negative in so far as it indicates the absence of a certain attribute, positive in so far as it implies that the thing which is declared to lack that attribute is of such a nature as to be capable of possessing it. A purely negative term conveys to the mind no positive information at all about the nature of the thing of which it is predicated, but leaves us to seek for it among the universe of things which fail to exhibit a given attribute.

A privative term, on the other hand, restricts us within a definite sphere. The term 'empty' restricts us within the sphere of things which are capable of fulness, that is, if the term be taken in its literal sense, things which possess extension in three dimensions.

§ 129. A positive and a negative term, which have the same matter, must exhaust the universe between them, e.g. 'white' and 'not-white,' since, according to the law of excluded middle, everything must be either one or the other. To say, however, that a thing is 'not-white' is merely to say that the term 'white' is inapplicable to it. 'Not-white' may be predicated of things which do not possess extension as well as of those which do. Such a pair of terms as 'white' and 'not-white,' in their relation to one another, are called Contradictories.

§ 130. Contrary terms must be distinguished from contradictory. Contrary terms are those which are most opposed under the same head. Thus 'white' and 'black' are contrary terms, being the most opposed under the same head of colour. 'Virtuous' and 'vicious' again are contraries, being the most opposed under the same head of moral quality.

§ 131. A positive and a privative term in the same matter will always be contraries, e.g. 'wise' and 'unwise,' 'safe' and 'unsafe': but contraries do not always assume the shape of positive and privative terms, but may both be positive in form, e.g. 'wise' and 'foolish,' 'safe' and 'dangerous.'

§ 132. Words which are positive in form are often privative in meaning, and vice versâ. This is the case, for instance, with the word 'safe,' which connotes nothing more than the absence of danger. We talk of a thing involving 'positive danger' and of its being 'positively unsafe' to do so and so. 'Unhappy,' on the other hand, signifies the presence of actual misery. Similarly in Latin 'inutilis' signifies not merely that there is no benefit to be derived from a thing, but that it is *positively injurious*. All such questions, however, are for the grammarian or lexicographer, and not for the logician. For the latter it is sufficient to know that corresponding to every term which signifies the presence of some attribute there may be imagined another which indicates the absence of the same attribute, where it might be possessed, and a third which indicates its absence, whether it might be possessed or not.

§ 133. Negative terms proper are formed by the prefix 'not-' or 'non-,' and are mere figments of logic. We do not in practice require to speak of the whole universe of objects minus those which possess a given attribute or collection of attributes. We have often occasion to speak of things which might be wise and are not, but seldom, if ever, of all things other than wise.

§ 134. Every privative attributive has, or may have, a corresponding abstract term, and the same is the case with negatives: for the absence of an attribute, is itself an attribute. Corresponding to 'empty,' there is 'emptiness'; corresponding to 'not-full' there may be imagined the term 'not-fulness.'

§ 135. The contrary of a given term always involves the contradictory, but it involves positive elements as well. Thus 'black' is 'not-white,' but it is something more besides. Terms which, without being directly contrary, involve a latent contradiction, are called Repugnant, e.g. 'red' and 'blue.' All terms whatever which signify attributes that exclude one another may be called Incompatible.

§ 136. The preceding division is based on what is known as the Quality of terms, a positive term being said to differ in quality from a non-positive one.

Univocal and Equivocal Terms.

§ 137. A term is said to be Univocal, when it has one and the same meaning wherever it occurs. A term which has more than one meaning is called Equivocal. 'Jam-pot,' 'hydrogen' are examples of univocal terms; 'pipe' and 'suit' of equivocal.

§ 138. This division does not properly come within the scope of logic, since it is a question of language, not of thought. From the logician's point of view an equivocal term is two or more different terms, for the definition in each sense would be different.

§ 139. Sometimes a third member is added to the same division under the head of Analogous Terms. The word 'sweet,' for instance, is applied by analogy to things so different in their own nature as a lump of sugar, a young lady, a tune, a poem, and so on. Again, because the head is the highest part of man, the highest part of a stream is called by analogy 'the head.' It is plainly inappropriate to make a separate class of analogous terms. Rather, terms become equivocal by being extended by analogy from one thing to another.

Absolute and Relative Terms.

§ 140. An Absolute term is a name given to a thing without reference to anything else.

§ 141. A Relative term is a name given to a thing with direct reference to some other thing.

§ 142. 'Hodge' and 'man' are absolute terms. 'Husband' 'father,' 'shepherd' are relative terms. 'Husband' conveys a direct reference to 'wife,' 'father' to 'Child,' 'shepherd' to 'sheep.' Given one term of a relation, the other is called the correlative, e.g. 'subject' is the correlative of 'ruler,' and conversely 'ruler' of 'subject.' The two terms are also spoken of as a pair of correlatives.

§ 143. The distinction between relative and absolute applies to attributives as well as subject-terms. 'Greater,' 'near,' 'like,' are instances of attributives which everyone would recognise as relative.

§ 144. A relation, it will be remembered, is a kind of attribute, differing from a quality in that it necessarily involves more substances than one. Every relation is at bottom a fact, or series of facts, in which two or more substances play a part. A relative term connotes this fact or facts from the point of view of one of the substances, its correlative from that of the other. Thus 'ruler' and 'subject' imply the same set of facts, looked at from opposite points of view. The series of facts itself, regarded from either side, is denoted by the corresponding abstract terms, 'rule 'and 'subjection.'

§ 145. It is a nice question whether the abstract names of relations should themselves be considered relative terms. Difficulties will perhaps be avoided by confining the expression 'relative *term*' to names of concrete things. 'Absolute,' it must be remembered, is a mere negative of 'relative,' and covers everything to which the definition of the latter does not strictly apply. Now it can hardly be said that 'rule' is a name given to a certain abstract thing with direct reference to some other thing, namely, subjection. Rather 'rule' and 'subjection' are two names for identically the same series of facts, according to the side from which we look at them. 'Ruler' and 'subject,' on the other hand, are names of two distinct substances, but each involving a reference to the other.

§ 146. This division then may be said to be based on the number of things involved in the name.

Connotative and Non-Connotative Terms.

§ 147. Before explaining this division, it is necessary to treat of what is called the Quantity of Terms.

Quantity of Terms.

§ 148. A term is possessed of quantity in two ways—

(1) In Extension;

(2) In Intension.

§ 149. The Extension of a term is the number of things to which it applies.

§ 150. The Intension of a term is the number of attributes which it implies.

§ 151. It will simplify matters to bear in mind that the intension of a term is the same thing as its meaning. To take an example, the term 'man' applies to certain things, namely, all the members of the human race that have been, are, or ever will be: this is its quantity in extension. But the term 'man' has also a certain meaning, and implies certain attributes—rationality, animality, and a definite bodily shape: the sum of these attributes constitutes its quantity in intension.

§ 152. The distinction between the two kinds of quantity possessed by a term is also conveyed by a variety of expressions which are here appended.

Extension = breadth = compass = application = denotation.

Intension = depth = comprehension = implication = connotation.

Of these various expressions, 'application' and 'implication' have the advantage of most clearly conveying their own meaning. 'Extension' and 'intension,' however, are more usual; and neither 'implication' nor 'connotation' is quite exact as a synonym for 'intension.' (§ 164.)

§ 153. We now return to the division of terms into connotative and non-connotative.

§ 154. A term is said to connote attributes, when it implies certain attributes at the same time that it applies to certain things distinct therefrom. [Footnote: Originally 'connotative' was used in the same sense in which we have used 'attributive,' for a word which directly signifies the presence of an attribute and indirectly applies to a subject. In this, its original sense, it was the subject which was said to be connoted, and not the attribute.]

§ 155. A term which possesses both extension and intension, distinct from one another, is connotative.

§ 156. A term which possesses no intension (if that be possible) or in which extension and intension coincide is non-connotative.

§ 157. The subject-term, 'man,' and its corresponding attributive, 'human,' have both extension and intension, distinct from one another. They are therefore connotative. But the abstract term, 'humanity,' denotes the very collection of attributes, which was before connoted by the concrete terms, 'man' and 'human.' In this case, therefore, extension and intension coincide, and the term is non-connotative.

§ 158. The above remark must be understood to be limited to abstract terms in their singular sense. When employed as common terms, abstract terms possess both extension and intension distinct from one another. Thus the term 'colour' applies to red, blue, and yellow, and at the same time implies (i.e. connotes), the power of affecting the eye.

§ 159. Since all terms are names of things, whether substances or attributes, it is clear that all terms must possess extension, though the extension of singular terms is the narrowest possible, as being confined to one thing.

§ 160. Are there then any terms which possess no intension? To ask this, is to ask—Are there any terms which have absolutely no meaning? It is often said that proper names are devoid of meaning, and the remark is, in a certain sense, true. When we call a being by the name 'man,' we do so because that being possesses human attributes, but when we call the same being by the name, 'John,' we do not mean to indicate the presence of any Johannine attributes. We simply wish to distinguish that being, in thought and language, from other beings of the same kind. Roughly speaking, therefore, proper names are devoid of meaning or intension. But no name can be entirely devoid of meaning. For, even setting aside the fact, which is not universally true, that proper names indicate the sex of the owner, the mere act of giving a name to a thing implies at least that the thing exists, whether in fact or thought; it implies what we may call 'thinghood': so that every term must carry with it some small amount of intension.

§ 161. From another point of view, however, proper names possess more intension than any other terms. For when we know a person, his name calls up to our minds all the individual attributes with which we are familiar, and these must be far more numerous than the attributes which are conveyed by any common term which can be applied to him. Thus the name 'John' means more to a person who knows him than 'attorney,' 'conservative,' 'scamp,' of 'vestry-man,' or any other term which may happen to apply to him. This, however, is the acquired intension of a term, and must be distinguished from the original intension. The name 'John' was never meant to indicate the attributes which its owner has, as a matter of fact, developed. He would be John all the same, if he were none of these.

§ 162. Hitherto we have been speaking only of christening-names, but it is evident that family names have a certain amount of connotation from the first. For when we dub John with the additional appellation of Smith, we do not give this second name as a mere individual mark, but intend thereby to indicate a relationship to other persons. The amount of connotation that can be conveyed by proper names is very noticeable in the Latin language. Let us take for an example the full name of a distinguished Roman— Publius Cornelius Scipio Æmilianus Africanus minor. Here it is only the prænomen, Publius, that can be said to be a mere individual mark, and even this distinctly indicates the sex of the owner. The nomen proper, Cornelius, declares the wearer of it to belong to the illustrious gens Cornelia. The cognomen, Scipio, further specifies him as a member of a distinguished family in that gens. The agnomen adoptivum indicates his transference by adoption from one gens to another. The second agnomen recalls the fact of his victory over the Carthaginians, while the addition of the word 'minor' distinguishes him from the former wearer of the same title. The name, instead of being devoid of meaning, is a chapter of history in itself. Homeric epithets, such as 'The Cloud-compeller,' 'The Earth-shaker' are instances of intensive proper names. Many of our own family names are obviously connotative in their origin, implying either some personal peculiarity, e.g. Armstrong, Cruikshank, Courteney; or the employment, trade or calling of the original bearer of the name, Smith, Carpenter, Baker, Clark, Leach, Archer, and so on; or else his abode, domain or nationality, as De Caen, De Montmorency, French, Langley; or simply the fact of descent from some presumably more noteworthy parent, as Jackson, Thomson, Fitzgerald, O'Connor, Macdonald, Apjohn, Price, Davids, etc. The question, however, whether a term is connotative or not, has to be decided, not by its origin, but by its use. We have seen that there are some proper names which, in a rough sense, may be said to possess no intension.

§ 163. The other kind of singular terms, namely, designations (§ 113) are obviously connotative. We cannot employ even the simplest of them without conveying more or less information about the qualities of the thing which they are used to denote. When, for instance, we say 'this table,' 'this book,' we indicate the proximity to the speaker of the object in question. Other designations have a higher degree of intension, as when we say 'the present prime minister of England,' 'the honourable member who brought forward this motion to-night.' Such terms have a good deal of significance in themselves, apart from any knowledge we may happen to possess of the individuals they denote.

§ 164. We have seen that, speaking quite strictly, there are no terms which are non-connotative: but, for practical purposes, we may apply the expression to proper names, on the ground that they possess no intension, and to singular abstract terms on the ground that their extension and intension coincide. In the latter case it is indifferent whether we call the quantity extension or intension. Only we cannot call it 'connotation,' because that implies two quantities distinct from one another. A term must already denote a subject before it can be said to connote its attributes.

§ 165. The division of terms into connotative and non-connotative is based on their possession of one quantity or two.

CHAPTER IV.

Of the Law of Inverse Variation of Extension and Intension.

§ 166. In a series of terms which fall under one another, as the extension decreases, the intension increases, and vice versâ. Take for instance the following series—

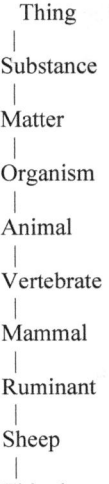

Thing
|
Substance
|
Matter
|
Organism
|
Animal
|
Vertebrate
|
Mammal
|
Ruminant
|
Sheep
|
This sheep.

Here the term at the top possesses the widest possible extension, since it applies to everything. But at the same time it possesses the least possible amount of intension, implying nothing more than mere existence, whether in fact or thought. On the other hand, the term at the bottom possesses the greatest amount of intension, since it implies all the attributes of, an individual superadded to those of the class to which it belongs: but its extension is the narrowest possible, being limited to one thing.

§ 167. At each step in the descent from the term at the top, which is called the 'Summum genus,' to the individual, we decrease the extension by increasing the intension. Thus by adding on to the bare notion of a thing the idea of independent existence, we descend to the term 'substance,' This process is known as Determination, or Specialisation.

§ 168. Again, by withdrawing our attention from the individual characteristics of a particular sheep, and fixing it upon those which are common to it with other animals of the same kind, we arrive at the common term, 'sheep.' Here we have increased the extension by decreasing the intension. This process is known as Generalisation.

§ 169. Generalisation implies abstraction, but we may have abstraction without generalisation.

§ 170. The following example is useful, as illustrating to the eye how a decrease of extension is accompanied by an increase of intension. At each step of the descent here we visibly tack on a fresh attribute. [Footnote: This example is borrowed from Professor Jevons.]

Ship
|
Steam-ship
|
Screw steam-ship

23

|
Iron screw steam-ship

|
British iron screw steam-ship.

Could we see the classes denoted by the names the pyramid would be exactly inverted.

§ 171. The law of inverse variation of extension and intension must of course be confined to the inter-relations of a series of terms of which each can be predicated of the other until we arrive at the bottom of the scale. It is not meant to apply to the extension and intension of the same term. The increase of population does not add to the meaning of 'baby.'

PART II.—OF PROPOSITIONS.

CHAPTER I.

Of the Proposition as distinguished from Other Sentences.

§ 172. As in considering the term, we found occasion to distinguish it from words generally, so now, in considering the proposition, it will be well to begin by distinguishing it from other sentences.

§ 173. Every proposition is a sentence, but every sentence is not a proposition.

§ 174. The field of logic is far from being conterminous with that of language. Language is the mirror of man's whole nature, whereas logic deals with language only so far as it gives clothing to the products of thought in the narrow sense which we have assigned to that term. Language has materials of every sort lying strewn about, among which the logician has to seek for his proper implements.

§ 175. Sentences may be employed for a variety of purposes—
(1) To ask a question;
(2) To give an order;
(3) To express a feeling;
(4) To make a statement.
These various uses give rise respectively to
(1) The Interrogative Sentence;
(2) The Imperative Sentence;
(3) The Exclamatory Sentence;
(4) The Enunciative Sentence; Indicative Potential.
It is with the last of these only that logic is concerned.

§ 176. The proposition, therefore, corresponds to the Indicative and Potential, or Conditional, sentences of grammar. For it must be borne in mind that logic recognises no difference between a statement of fact and a supposition. 'It may rain to-morrow' is as much a proposition as 'It is raining now.'

§ 177. Leaving the grammatical aspect of the proposition, we must now consider it from the purely logical point of view.

§ 178. A proposition is a judgement expressed in words; and a judgement is a direct comparison between two concepts.

§ 179. The same thing may be expressed more briefly by saying that a proposition is a direct comparison between two terms.

§ 180. We say 'direct comparison,' because the syllogism also may be described as a comparison between two terms: but in the syllogism the two terms are compared indirectly, or by means of a third term.

§ 181. A proposition may be analysed into two terms and a Copula, which is nothing more than the sign of agreement or disagreement between them.

§ 182. The two terms are called the Subject and the Predicate (§ 58).

§ 183. The Subject is that of which something is stated.

§ 184. The Predicate is that which is stated of the subject.

§ 185. Hence the subject is thought of for its own sake, and the predicate for the sake of the subject.

CHAPTER II.

Of *the Copula.*

§ 186. There are two kinds of copula, one for affirmative and one for negative statements.

§ 187. Materially the copula is expressed by some part of the verb 'to be,' with or without the negative, or else is wrapped up in some inflexional form of a verb.

§ 188. The material form of the copula is an accident of language, and a matter of indifference to logic. 'The kettle boils' is as logical a form of expression as 'The kettle is boiling.' For it must be remembered that the word 'is' here is a mere sign of agreement between the two terms, and conveys no notion of actual existence. We may use it indeed with equal propriety to express non-existence, as when we say 'An idol is nothing.'

§ 189. When the verb 'to be' expresses existence in fact it is known in grammar as 'the substantive verb.' In this use it is predicate as well as copula, as when we say 'God is,' which may be analysed, if we please, into 'God is existent.'

§ 190. We have laid down above that there are two kinds of copula, affirmative and negative: but some logicians have maintained that the copula is always affirmative.

§ 191. What then, it may be asked, on this view, is the meaning of negative propositions! To which the answer is, that a negative proposition asserts an agreement between the subject and a negative term. When, for instance, we say 'The whale is not a fish,' this would be interpreted to mean 'The whale is a not-fish.'

§ 192. Undoubtedly any negative proposition may be exhibited in an affirmative form, since, by the law of excluded middle, given a pair of contradictory terms, wherever the one can be asserted, the other can be denied, and vice versâ. We shall find later on that this principle gives rise to one of the forms of immediate inference. The only question then can be, which is the more natural and legitimate form of expression. It seems simpler to suppose that we assert the agreement of 'whale' with 'not-fish' by implication only, and that what we directly do is to predicate a disagreement between 'whale' and the positive attributes connoted by 'fish.' For since 'not-fish' must apply to every conceivable object of thought except those which fall under the positive term 'fish,' to say that a whale is a 'not-fish,' is to say that we have still to search for 'whale' throughout the whole universe of being, minus a limited portion; which is only a more clumsy way of saying that it is not to be found in that portion.

§ 193. Again, the term 'not-fish' must be understood either in its intension or in its extension. If it be understood in its intension, what it connotes is simply the absence of the positive qualities which constitute a fish, a meaning which is equally conveyed by the negative form of proposition. We gain nothing in simplicity by thus confounding assertion with denial. If, on the other hand, it is to be taken in extension, this involves the awkwardness of supposing that the predicative power of a term resides in its extensive capacity.

§ 194. We therefore recognise predication as being of two kinds—affirmation and negation—corresponding to which there are two forms of copula.

§ 195. On the other hand, other logicians have maintained that there are many kinds of copula, since the copula must vary according to the various degrees of probability with which we can assert or deny a predicate of a subject. This view is technically known as the doctrine of

The Modality of the Copula.

§ 196. It may plausibly be maintained that the division of propositions into affirmative and negative is not an exhaustive one, since the result of an act of judgement is not always to lead the mind to a clear assertion or a clear denial, but to leave it in more or less doubt as to whether the predicate applies to the subject or not. Instead of saying simply A is B, or A is not B, we may be led to one of the following forms of proposition—

A is possibly B.

A is probably B.

A is certainly B.

The adverbial expression which thus appears to qualify the copula is known as 'the mode.'

§ 197. When we say 'The accused may be guilty' we have a proposition of very different force from 'The accused is guilty,' and yet the terms appear to be the same. Wherein then does the difference lie? 'In the copula' would seem to be the obvious reply. We seem therefore driven to admit that there are as many different kinds of copula as there are different degrees of assurance with which a statement may be made.

§ 198. But there is another way in which modal propositions may be regarded. Instead of the mode being attached to the copula, it may be considered as itself constituting the predicate, so that the above propositions would be analysed thus—

That A is B, is possible.

That A is B, is probable.

That A is B, is certain.

§ 199. The subject here is itself a proposition of which we predicate various degrees of probability. In this way the division of propositions into affirmative and negative is rendered exhaustive. For wherever before we had a doubtful assertion, we have now an assertion of doubtfulness.

§ 200. If degrees of probability can thus be eliminated from the copula, much more so can expressions of time, which may always be regarded as forming part of the predicate. 'The sun will rise to-morrow' may be analysed into 'The sun is going to rise to-morrow.' In either case the tense belongs equally to the predicate. It is often an awkward task so to analyse propositions relative to past or future time as to bring out the copula under the form 'is' or 'is not': but fortunately there is no necessity for so doing, since, as has been said before (§ 188), the material form of the copula is a matter of indifference to logic. Indeed in affirmative propositions the mere juxtaposition of the subject and predicate is often sufficient to indicate their agreement, e.g. 'Most haste, worst speed,' chalepha tha kala. It is because all propositions are not affirmative that we require a copula at all. Moreover the awkwardness of expression just alluded to is a mere accident of language. In Latin we may say with equal propriety 'Sol orietur cras' or 'Sol est oriturus cras'; while past time may also be expressed in the analytic form in the case of deponent verbs, as 'Caesar est in Galliam profectus'—'Caesar is gone into Gaul.'

§ 201. The copula then may always be regarded as pure, that is, as indicating mere agreement or disagreement between the two terms of the proposition.

CHAPTER III.

Of the Divisions of Propositions.

§ 202. The most obvious and the most important division of propositions is into true and false, but with this we are not concerned. Formal logic can recognise no difference between true and false propositions. The one is represented by the same symbols as the other.

§ 203. We may notice, however, in passing, that truth and falsehood are attributes of propositions and of propositions only. For something must be predicated, i.e. asserted or denied, before we can have either truth or falsehood. Neither concepts or terms, on the one hand, nor reasonings, on the other, can properly be said to be true or false. In the mere notion of a Centaur or of a black swan there is neither truth nor falsehood; it is not until we make some statement about these things, such as that 'black swans are found in Australia,' or 'I met a Centaur in the High Street yesterday,' that the question of truth or falsehood comes in. In such expressions as a 'true friend' or 'a false patriot' there is a tacit reference to propositions. We mean persons of whom the terms 'friend' and 'patriot' are truly or falsely predicated. Neither can we with any propriety talk of true or false reasoning. Reasoning is either valid or invalid: it is only the premises of our reasonings, which are propositions, that can be true or false. We may have a perfectly valid process of reasoning which starts from a false assumption and lands us in a false conclusion.

§ 204. All truth and falsehood then are contained in propositions; and propositions are divided according to the Quality of the Matter into true and false. But the consideration of the matter is outside the sphere of formal or deductive Logic. It is the problem of inductive logic to establish, if possible, a criterion of evidence whereby the truth or falsehood of propositions may be judged (§ 2).

§ 205. Another usual division of propositions is into Pure and Modal, the latter being those in which the copula is modified by some degree of probability. This division is excluded by the view which has just been taken of the copula, as being always simply affirmative or simply negative.

§ 206. We are left then with the following divisions of propositions—

Proposition
according to Form
Simple
Complex
Conjunctive
Disjunctive
Universal
Singular
General
according to Matter
Verbal
Real
according to Quantity
Universal

Singular
General
    Particular
    Indefinite
(strictly) Particular
    according to Quality
    Affirmative
    Negative

Simple and Complex Propositions.

§ 207. A Simple Proposition is one in which a predicate is directly affirmed or denied of a subject, e.g. 'Rain is falling.'

§ 208. A simple proposition is otherwise known as Categorical.

§ 209. A Complex Proposition is one in which a statement is made subject to some condition, e.g. 'If the wind drops, rain will fall.'

§ 210. Hence the complex proposition is also known as Conditional.

§ 211. Every complex proposition consists of two parts—

(1) Antecedent;

(2) Consequent.

§ 212. The Antecedent is the condition on which another statement is made to depend. It precedes the other in the order of thought, but may either precede or follow it in the order of language. Thus we may say indifferently—'If the wind drops, we shall have rain' or 'We shall have rain, if the wind drops.'

§ 213. The Consequent is the statement which is made subject to some condition.

§ 214. The complex proposition assumes two forms,

(1) If A is B, C is D.

This is known as the Conjunctive or Hypothetical proposition.

(2) Either A is B or C is D.

This is known as the Disjunctive proposition.

§ 215. The disjunctive proposition may also appear in the form

A is either B or C,

which is equivalent to saying

Either A is B or A is C;

or again in the form

Either A or B is C,

which is equivalent to saying

Either A is C or B is C.

§ 216. As the double nomenclature may cause some confusion, a scheme is appended.

        Proposition
```
  _____|_____
 | |
```
Simple Complex
(Categorical) (Conditional)
```
          _____|_____
         | |
```
       Conjunctive Disjunctive.
       (Hypothetical)

§ 217. The first set of names is preferable. 'Categorical' properly means 'predicable' and 'hypothetical' is a mere synonym for 'conditional.'

§ 218. Let us examine now what is the real nature of the statement which is made in the complex form of proposition. When, for instance, we say 'If the sky falls, we shall catch larks,' what is it that we really mean to assert? Not that the sky will fall, and not that we shall catch larks, but a certain connection between the two, namely, that the truth of the antecedent involves the truth of the consequent. This is why this form of proposition is called 'conjunctive,' because in it the truth of the consequent is conjoined to the truth of the antecedent.

§ 219. Again, when we say 'Jones is either a knave or a fool,' what is really meant to be asserted is—'If you do not find Jones to be a knave, you may be sure that he is a fool.' Here it is the falsity of the antecedent which involves the truth of the consequent; and the proposition is known as 'disjunctive,' because the truth of the consequent is disjoined from the truth of the antecedent.

§ 220. Complex propositions then turn out to be propositions about propositions, that is, of which the subject and predicate are themselves propositions. But the nature of a proposition never varies in thought. Ultimately every proposition must assume the form 'A is, or is not, B.' 'If the sky falls, we shall catch larks' may be compressed into 'Sky-falling is lark-catching.'

§ 221. Hence this division turns upon the form of expression, and may be said to be founded on the simplicity or complexity of the terms employed in a proposition.

§ 222. In the complex proposition there appears to be more than one subject or predicate or both, but in reality there is only a single statement; and this statement refers, as we have Seen, to a certain connection between two propositions.

§ 223. If there were logically, and not merely grammatically, more than one subject or predicate, there would be more than one proposition. Thus when we say 'The Jews and Carthaginians were Semitic peoples and spoke a Semitic language,' we have four propositions compressed into a single sentence for the sake of brevity.

§ 224. On the other hand when we say 'Either the Carthaginians were of Semitic origin or argument from language is of no value in ethnology,' we have two propositions only in appearance.

§ 225. The complex proposition then must be distinguished from those contrivances of language for abbreviating expression in which several distinct statements are combined into a single sentence.

Verbal and Real Propositions.

§ 226. A Verbal Proposition is one which states nothing more about the subject than is contained in its definition, e.g. 'Man is an animal'; 'Men are rational beings.'

§ 227. A Real Proposition states some fact not contained in the definition of the subject, e.g. 'Some animals have four feet.'

§ 228. It will be seen that the distinction between verbal and real propositions assumes a knowledge of the precise meaning of terms, that is to say, a knowledge of definitions.

§ 229. To a person who does not know the meaning of terms a verbal proposition will convey as much information as a real one. To say 'The sun is in mid-heaven at noon,' though a merely verbal proposition, will convey information to a person who is being taught to attach a meaning to the word 'noon.' We use so many terms without knowing their meaning, that a merely verbal proposition appears a revelation to many minds. Thus there are people who are surprised to hear that the lion is a cat, though in its definition 'lion' is referred to the class 'cat.' The reason of this is that we know material objects far better in their extension than in their intension, that is to say, we know what things a name applies to without knowing the attributes which those things possess in common.

§ 230. There is nothing in the mere look of a proposition to inform us whether it is verbal or real; the difference is wholly relative to, and constituted by, the definition of the subject. When we have accepted as the definition of a triangle that it is 'a figure contained by three sides,' the statement of the further fact that it has three angles becomes a real proposition. Again the proposition 'Man is progressive' is a real proposition. For though his progressiveness is a consequence of his rationality, still there is no actual reference to progressiveness contained in the usually accepted definition, 'Man is a rational animal.'

§ 231. If we were to admit, under the term 'verbal proposition,' all statements which, though not actually contained in the definition of the subject, are implied by it, the whole body of necessary truth would have to be pronounced merely verbal, and the most penetrating conclusions of mathematicians set down as only another way of stating the simplest axioms from which they started. For the propositions of which necessary truth is composed are so linked together that, given one, the rest can always follow. But necessary truth, which is arrived at 'a priori,' that is, by the mind's own working, is quite as real as contingent truth, which is arrived at 'a posteriori,' or by the teachings of experience, in other words, through our own senses or those of others.

§ 232. The process by which real truth, which is other than deductive, is arrived at 'a priori' is known as Intuition. E.g. The mind sees that what has three sides cannot but have three angles.

§ 233. Only such propositions then must be considered verbal as state facts expressly mentioned in the definition.

§ 234. Strictly speaking, the division of propositions into verbal and real is extraneous to our subject: since it is not the province of logic to acquaint us with the content of definitions.

§ 235, The same distinction as between verbal and real proposition, is conveyed by the expressions 'Analytical' and 'Synthetical,' or 'Explicative' and 'Ampliative' judgements.

§ 236. A verbal proposition is called analytical, as breaking up the subject into its component notions.

§ 237. A real proposition is called synthetical, as attaching some new notion to the subject.

§ 238. Among the scholastic logicians verbal propositions were known as 'Essential,' because what was stated in the definition was considered to be of the essence of the subject, while real propositions were known as 'Accidental.'

Universal AND PARTICULAR Propositions.

§ 239. A Universal proposition is one in which it is evident from the form that the predicate applies to the subject in its whole extent.

§ 240. When the predicate does not apply to the subject in its whole extent, or when it is not clear that it does so, the proposition is called Particular.

§ 241. To say that a predicate applies to a subject in its whole extent, is to say that it is asserted or denied of all the things of which the subject is a name.

§ 242. 'All men are mortal' is a universal proposition.

§ 243. 'Some men are black' is a particular proposition. So also is 'Men are fallible;' for here it is not clear from the form whether 'all' or only 'some' is meant.

§ 244. The latter kind of proposition is known as Indefinite, and must be distinguished from the particular proposition strictly so called, in which the predicate applies to part only of the subject.

§ 245. The division into universal and particular is founded on the Quantity of propositions.

§ 246. The quantity of a proposition is determined by the quantity in extension of its subject.

§ 247. Very often the matter of an indefinite proposition is such as clearly to indicate to us its quantity. When, for instance, we say 'Metals are elements,' we are understood to be referring to all metals; and the same thing holds true of scientific statements in general. Formal logic, however, cannot take account of the matter of propositions; and is therefore obliged to set down all indefinite propositions as particular, since it is not evident from the form that they are universal.

§ 248. Particular propositions, therefore, are sub-divided into such as are Indefinite and such as are Particular, in the strict sense of the term.

§ 249. We must now examine the sub-division of universal propositions into Singular and General.

§ 250. A Singular proposition is one which has a singular term for its subject, e.g. 'Virtue is beautiful.'

§ 251. A General proposition is one which has for its subject a common term taken in its whole extent.

§ 252. Now when we say 'John is a man' or 'This table is oblong,' the proposition is quite as universal, in the sense of the predicate applying to the whole of the subject, as when we say 'All men are mortal.' For since a singular term applies only to one thing, we cannot avoid using it in its whole extent, if we use it at all.

§ 253. The most usual signs of generality in a proposition are the words 'all,' 'every,' 'each,' in affirmative, and the words 'no,' 'none,' 'not one,' &c. in negative propositions.

§ 254. The terminology of the division of propositions according to quantity is unsatisfactory. Not only has the indefinite proposition to be set down as particular, even when the sense manifestly declares it to be universal; but the proposition which is expressed in a particular form has also to be construed as indefinite, *so* that an unnatural meaning is imparted to the word 'some,' as used in logic. If in common conversation we were to say 'Some cows chew the cud,' the person whom we were addressing would doubtless imagine us to suppose that there were some cows which did not possess this attribute. But in logic the word 'some' is not held to express more than 'some at least, if not all.' Hence we find not only that an indefinite proposition may, as a matter of fact, be strictly particular, but that a proposition which appears to be strictly particular may be indefinite. So a proposition expressed in precisely the same form 'Some A is B' may be either strictly particular, if some be taken to exclude all, or indefinite, if the word 'some' does not exclude the possibility of the statement being true of all. It is evident that the term 'particular' has become distorted from its original meaning. It would naturally lead us to infer that a statement is limited to part of the subject, whereas, by its being opposed to universal, in the sense in which that term has been defined, it can only mean that we have nothing to show us whether part or the whole is spoken of.

§ 255. This awkwardness of expression is due to the indefinite proposition having been displaced from its proper position. Formerly propositions were divided under three heads—

(1) Universal,
(2) Particular,
(3) Indefinite.

But logicians anxious for simplification asked, whether a predicate in any given case must not either apply to the whole of the subject or not? And whether, therefore, the third head of indefinite propositions were not as superfluous as the so-called 'common gender' of nouns in grammar?

§ 256. It is quite true that, as a matter of fact, any given predicate must either apply to the whole of the subject or not, so that in the nature of things there is no middle course between universal and particular. But the important point is that we may not know whether the predicate applies to the whole of the subject or not. The primary division then should be into propositions whose quantity is known and propositions whose quantity is unknown. Those propositions whose quantity is known may be sub-divided into 'definitely universal' and 'definitely particular,' while all those whose quantity is unknown are classed together under the term 'indefinite.' Hence the proper division is as follows—

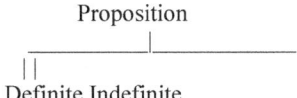

Proposition

Definite Indefinite

```
_____|_____
| |
```

Universal Particular.

§ 257. Another very obvious defeat of terminology is that the word 'universal' is naturally opposed to 'singular,' whereas it is here so used as to include it; while, on the other hand, there is no obvious difference between universal and general, though in the division the latter is distinguished from the former as species from genus.

Affirmative and Negative Propositions.

§ 258. This division rests upon the Quality of propositions.

§ 259. It is the quality of the form to be affirmative or negative: the quality of the matter, as we saw before (§ 204), is to be true or false. But since formal logic takes no account of the matter of thought, when we speak of 'quality' we are understood to mean the quality of the form.

§ 260. By combining the division of propositions according to quantity with the division according to quality, we obtain four kinds of proposition, namely—

(1) Universal Affirmative (A).

(2) Universal Negative (E).

(3) Particular Affirmative (I).

(4) Particular Negative (O).

§ 261. This is an exhaustive classification of propositions, and any proposition, no matter what its form may be, must fall under one or other of these four heads. For every proposition must be either universal or particular, in the sense that the subject must either be known to be used in its whole extent or not; and any proposition, whether universal or particular, must be either affirmative or negative, for by denying modality to the copula we have excluded everything intermediate between downright assertion and denial. This classification therefore may be regarded as a Procrustes' bed, into which every proposition is bound to fit at its proper peril.

§ 262. These four kinds of propositions are represented respectively by the symbols A, E, I, O.

§ 263. The vowels A and I, which denote the two affirmatives, occur in the Latin words 'affirmo' and 'aio;' E and O, which denote the two negatives, occur in the Latin word 'nego.'

Extensive and Intensive Propositions.

§ 264. It is important to notice the difference between Extensive and Intensive propositions; but this is not a division of propositions, but a distinction as to our way of regarding them. Propositions may be read either in extension or intension. Thus when we say 'All cows are ruminants,' we may mean that the class, cow, is contained in the larger class, ruminant. This is reading the proposition in extension. Or we may mean that the attribute of chewing the cud is contained in, or accompanies, the attributes which make up our idea of 'cow.' This is reading the proposition in intension. What, as a matter of fact, we do mean, is a mixture of the two, namely, that the class, cow, has the attribute of chewing the cud. For in the ordinary and natural form of proposition the subject is used in extension, and the predicate in intension, that is to say, when we use a subject, we are thinking of certain objects, whereas when we use a predicate, we indicate the possession of certain attributes. The predicate, however, need not always be used in intension, e.g. in the proposition 'His name is John' the predicate is not intended to convey the idea of any attributes at all. What is meant to be asserted is that the name of the person in question is that particular name, John, and not Zacharias or Abinadab or any other name that might be given him.

§ 265. Let it be noticed that when a proposition is read in extension, the predicate contains the subject, whereas, when it is read in intension, the subject contains the predicate.

Exclusive Propositions.

§ 266. An Exclusive Proposition is so called because in it all but a given subject is excluded from participation in a given predicate, e.g. 'The good alone are happy,' 'None but the brave deserve the fair,' 'No one except yourself would have done this.'

§ 267. By the above forms of expression the predicate is declared to apply to a given subject and to that subject only. Hence an exclusive proposition is really equivalent to two propositions, one affirmative and one negative. The first of the above propositions, for instance, means that some of the good are happy, and that no one else is so. It does not necessarily mean that all the good are happy, but asserts that among the good will be found all the happy. It is therefore equivalent to saying that all the happy are good, only that it puts prominently forward in addition what is otherwise a latent consequence of that assertion, namely, that some at least of the good are happy.

§ 268. Logically expressed the exclusive proposition when universal assumes the form of an E proposition, with a negative term for its subject

No not-A is B.

§ 269. Under the head of exclusive comes the strictly particular proposition, 'Some A is B,' which implies at the same time that 'Some A is not B.' Here 'some' is understood to mean 'some only,' which is the meaning that it usually bears in common language. When, for instance, we say 'Some of the gates into the park are closed at nightfall,' we are understood to mean 'Some are left open.'

Exceptive Propositions.

§ 270. An Exceptive Proposition is so called as affirming the predicate of the whole of the subject, with the exception of a certain part, e.g. 'All the jury, except two, condemned the prisoner.'

§ 271. This form of proposition again involves two distinct statements, one negative and one affirmative, being equivalent to 'Two of the jury did not condemn the prisoner; and all the rest did.'

§ 272. The exceptive proposition is merely an affirmative way of stating the exclusive—

No not-A is B = All not-A is not-B.

No one but the sage is sane = All except the sage are mad.

Tautologous or Identical Propositions

§ 273. A Tautologous or Identical proposition affirms the subject of itself, e.g. 'A man's a man,' 'What I have written, I have written,' 'Whatever is, is.' The second of these instances amounts formally to saying 'The thing that I have written is the thing that I have written,' though of course the implication is that the writing will not be altered.

CHAPTER IV.

Of the Distribution of Terms.

§ 274. The treatment of this subject falls under the second part of logic, since distribution is not an attribute of terms in themselves, but one which they acquire in predication.

§ 275. A term is said to be distributed when it is known to be used in its whole extent, that is, with reference to all the things of which it is a name. When it is not so used, or is not known to be so used, it is called undistributed.

§ 276. When we say 'All men are mortal,' the subject is distributed, since it is apparent from the form of the expression that it is used in its whole extent. But when we say 'Men are miserable' or 'Some men are black,' the subject is undistributed.

§ 277. There is the same ambiguity attaching to the term 'undistributed' which we found to underlie the use of the term 'particular.' 'Undistributed' is applied both to a term whose quantity is undefined, and to one whose quantity is definitely limited to a part of its possible extent.

§ 278. This awkwardness arises from not inquiring first whether the quantity of a term is determined or undetermined, and afterwards proceeding to inquire, whether it is determined as a whole or part of its possible extent. As it is, to say that a term is distributed, involves two distinct statements—

(1) That its quantity is known;

(2) That its quantity is the greatest possible.

The term 'undistributed' serves sometimes to contradict one of these statements and sometimes to contradict the other.

§ 279. With regard to the quantity of the subject of a proposition no difficulty can arise. The use of the words 'all' or 'some,' or of a variety of equivalent expressions, mark the subject as being distributed or undistributed respectively, while, if there be nothing to mark the quantity, the subject is for that reason reckoned undistributed.

§ 280. With regard to the predicate more difficulty may arise.

§ 281. It has been laid down already that, in the ordinary form of proposition, the subject is used in extension and the predicate in intension. Let us illustrate the meaning of this by an example. If someone were to say 'Cows are ruminants,' you would have a right to ask him whether he meant 'all cows' or only 'some.' You would not by so doing be asking for fresh information, but merely for a more distinct explanation of the statement already made. The subject being used in extension naturally assumes the form of the whole or part of a class. But, if you were to ask the same person 'Do you mean that cows are all the ruminants that there are, or only some of them?' he would have a right to complain of the question, and might fairly reply, 'I did not mean either one or the other; I was not thinking of ruminants as a class. I wished merely to assert an attribute of cows; in fact, I meant no more than that cows chew the cud.'

§ 282. Since therefore a predicate is not used in extension at all, it cannot possibly be known whether it is used in its whole extent or not.

§ 283. It would appear then that every predicate is necessarily undistributed; and this consequence does follow in the case of affirmative propositions.

§ 284. In a negative proposition, however, the predicate, though still used in intension, must be regarded as distributed. This arises from the nature of a negative proposition. For we must remember that in any proposition, although the predicate be not meant in extension, it always admits of being so read. Now we cannot exclude one class from another without at the same time wholly excluding that other from the former. To take an example, when we say 'No horses are ruminants,' the meaning we really wish to convey is that no member of the class, horse, has a particular attribute, namely, that of chewing the cud. But the proposition admits of being read in another form, namely, 'That no member of the class, horse, is a member of the class, ruminant.' For by excluding a class from the possession of a given attribute, we inevitably exclude at the same time any class of things which possess that attribute from the former class.

§ 285. The difference between the use of a predicate in an affirmative and in a negative proposition may be illustrated to the eye as follows. To say 'All A is B' may mean either that A is included in B or that A and B are exactly co-extensive.

[Illustration]

34

§ 286. As we cannot be sure which of these two relations of A to B is meant, the predicate B has to be reckoned undistributed, since a term is held to be distributed only when we know that it is used in its whole extent.

§ 287. To say 'No A is B,' however, is to say that A falls wholly outside of B, which involves the consequence that B falls wholly outside of A.

[Illustration]

§ 288. Let us now apply the same mode of illustration to the particular forms of proposition.

§ 289. If I be taken in the strictly particular sense, there are, from the point of view of extension, two things which may be meant when we say 'Some A is B'—

(1) That A and B are two classes which overlap one another, that is to say, have some members in common, e.g. 'Some cats are black.'

[Illustration]

(2) That B is wholly contained in A, which is an inverted way of saying that all B is A, e.g. 'Some animals are men.'

[Illustration]

§ 290. Since we cannot be sure which of these two is meant, the predicate is again reckoned undistributed.

§ 291. If on the other hand 1 be taken in an indefinite sense, so as to admit the possibility of the universal being true, then the two diagrams which have already been used for A must be extended to 1, in addition to its own, together with the remarks which we made in connection with them (§§ 285-6).

§ 292. Again, when we say 'Some A is not B,' we mean that some, if not the whole of A, is excluded from the possession of the attribute B. In either case the things which possess the attribute B are wholly excluded either from a particular part or from the whole of A. The predicate therefore is distributed.

[Illustration]

From the above considerations we elicit the following—

§ 293. Four Rules for the Distribution of Terms.

(1) All universal propositions distribute their subject.

(2) No particular propositions distribute their subject,

(3) All negative propositions distribute their predicate.

(4) No affirmative propositions distribute their predicate.

§ 294. The question of the distribution or non-distribution of the subject turns upon the quantity of the proposition, whether universal or particular; the question of the distribution or non-distribution of the predicate turns upon the quality of the proposition, whether affirmative or negative.

CHAPTER V.

Of the Quantification of the Predicate.

§ 295. The rules that have been given for the distribution of terms, together with the fourfold division of propositions into A, E, 1, 0, are based on the assumption that it is the distribution or non-distribution of the subject only that needs to be taken into account in estimating the quantity of a proposition.

§ 296. But some logicians have maintained that the predicate, though seldom quantified in expression, must always be quantified in thought—in other words, that when we say, for instance, 'All A is B,' we must mean either that 'All A is all B' or only that 'All A is some B.'

§ 297. If this were so, it is plain that the number of possible propositions would be exactly doubled, and that, instead of four forms, we should now have to recognise eight, which may be expressed as follows—

1. All A is all B. ([upsilon]).
2. All A is some B. ([Lambda]).
3. No A is any B. ([Epsilon]).
4. No A is some B. ([eta]).
5. Some A is all B. ([Upsilon]).
6. Some A is some B. ([Iota]).
7. Some A is not any B. ([Omega]).
8. Some A is not some B. ([omega]).

§ 298. It is evident that it is the second of the above propositions which represents the original A, in accordance with the rule that 'No affirmative propositions distribute their predicate' (§ 293).

§ 299. The third represents the original E, in accordance with the rule that 'All negative propositions distribute their predicate.'

§ 300. The sixth represents the original I, in accordance with the rule that 'No affirmative propositions distribute their predicate.'

§ 301. The seventh represents the original O, in accordance with the rule that 'All negative propositions distribute their predicate.'

§ 302. Four new symbols are required, if the quantity of the predicate as well as that of the subject be taken into account in the classification of propositions. These have been supplied, somewhat fancifully, as follows—

§ 303. The first, 'All A is all B,' which distributes both subject and predicate, has been called [upsilon], to mark its extreme universality.

§ 304. The fourth, 'No A is some B,' is contained in E, and has therefore been denoted by the symbol [eta], to show its connection with E.

§ 305. The fifth, 'Some A is all B,' is the exact converse of the second, 'All A is some B,' and has therefore been denoted by the symbol [Upsilon], which resembles an inverted A.

§ 306. The eighth is contained in O, as part in whole, and has therefore had assigned to it the symbol [omega],

§ 307. The attempt to take the predicate in extension, instead of, as it should naturally be taken, in intension, leads to some curious results. Let us take, for instance, the u proposition. Either the sign of quantity 'all' must be understood as forming part of the predicate or not. If it is not, then the u proposition 'All A is all B' seems to contain within itself, not one proposition, but two, namely, 'All A is B' and 'All B is A.' But if on the other hand 'all' is understood to form part of the predicate, then u is not really a general but a singular proposition. When we say, 'All men are rational animals,' we have a true general proposition, because the predicate applies to the subject distributively, and not collectively. What we mean is that 'rational animal' may be affirmed of every individual in the class, man. But when we say 'All men are all rational animals,' the predicate no longer applies to the subject distributively, but only collectively. For it is obvious that 'all rational animals' cannot be affirmed of every individual in the class, man. What the proposition means is that the class, man, is co-extensive with the class, rational animal. The same meaning may be expressed intensively by saying that the one class has the attribute of co-extension with the other.

§ 308. Under the head o u come all propositions in which both subject and predicate are singular terms, e.g. 'Homer was the author of the Iliad,' 'Virtue is the way to happiness.'

§ 309. The proposition [eta] conveys very little information to the mind. 'No A is some B' is compatible with the A proposition in the same matter. 'No men are some animals' may be true, while at the same time it is true that 'All men are animals.' No men, for instance, are the particular animals known as kangaroos.

§ 310. The [omega] proposition conveys still less information than the [eta]. For [omega] is compatible, not only with A, but with [upsilon]. Even though 'All men are all rational animals,' it is still true that 'Some men are not some rational animals': for no given human being is the same rational animal as any other.

§ 311. Nay, even when the [upsilon] is an identical proposition, [omega] will still hold in the same matter. 'All rational animals are all rational animals': but, for all that, 'Some rational animals are not some others.' This last form of proposition therefore is almost wholly devoid of meaning.

§ 312. The chief advantage claimed for the quantification of the predicate is that it reduces every affirmative proposition to an exact equation between its subject and predicate. As a consequence every proposition would admit of simple conversion, that is to say, of having the subject and predicate transposed without any further change in the proposition. The forms also of Reduction (a term which will be explained later on) would be simplified; and generally the introduction of the quantified predicate into logic might be attended with certain mechanical advantages. The object of the logician, however, is not to invent an ingenious system, but to arrive at a true analysis of thought. Now, if it be admitted that in the ordinary form of proposition the subject is used in extension and the predicate in intension, the ground for the doctrine is at once cut away. For, if the predicate be not used in its extensive capacity at all, we plainly cannot be called upon to determine whether it is used in its whole extent or not.

CHAPTER VI.

Of the Heads of Predicables.

§ 313. A predicate is something which is stated of a subject.

§ 314. A predicable is something which can be stated of a subject.

§ 315. The Heads of Predicables are a classification of the various things which can be stated of a subject, viewed in their relation to it.

§ 316. The treatment of this topic, therefore, as it involves the relation of a predicate to a subject, manifestly falls under the second part of logic, which deals with the proposition. It is sometimes treated under the first part of logic, as though the heads of predicables were a classification of universal notions, i.e. common terms, in relation to one another, without reference to their place in the proposition.

§ 317. The heads of predicables are commonly reckoned as five, namely,

(1) Genus.

(2) Species.

(3) Difference.

(4) Property.

(5) Accident.

§ 318. We will first define these terms in the sense in which they are now used, and afterwards examine the principle on which the classification is founded and the sense in which they were originally intended.

(1) A Genus is a larger class containing under it smaller classes. Animal is a genus in relation to man and brute.

(2) A Species is a smaller class contained under a larger one. Man is a species in relation to animal.

(3) Difference is the attribute, or attributes, which distinguish one species from others contained under the same genus. Rationality is the attribute which distinguishes the species, man, from the species, brute.

N.B. The genus and the difference together make up the Definition of a class-name, or common term.

(4) A Property is an attribute which is not contained in the definition of a term, but which flows from it.

A Generic Property is one which flows from the genus.

A Specific Property is one which flows from the difference.

It is a generic property of man that he is mortal, which is a consequence of his animality. It is a specific property of man that he is progressive, which is a consequence of his rationality.

(5) An Accident is an attribute, which is neither contained in the definition, nor flows from it.

§ 319. Accidents are either Separable or Inseparable.

A Separable Accident is one which belongs only to some members of a class.

An Inseparable Accident is one which belongs to all the members of a class.

Blackness is a separable accident of man, an inseparable accident of coals.

§ 320. The attributes which belong to anything may be distinguished broadly under the two heads of essential and non-essential, or accidental. By the essential attributes of anything are meant those which are contained in, or which flow from, the definition. Now it may be questioned whether there can, in the nature of things, be such a thing as an inseparable accident. For if an attribute were found to belong invariably to all the members of a class, we should suspect that there was some causal connection between it and the attributes which constitute the definition, that is, we should suspect the attribute in question to be essential and not accidental. Nevertheless the term 'inseparable accident' may be retained as a cloak for our ignorance, whenever it is found that an attribute does, as a matter of fact, belong to all the members of a class, without there being any apparent reason why it should do so. It has been observed that animals which have horns chew the cud. As no one can adduce any reason why animals that have horns should chew the cud any more than animals which have not, we may call the fact of chewing the cud an inseparable accident of horned animals.

§ 321. The distinction between separable and inseparable accidents is sometimes extended from classes to individuals.

An inseparable accident of an individual is one which belongs to him at all times. A separable accident of an individual is one which belongs to him at one time and not at another.

§ 322. It is an inseparable accident of an individual that he was born at a certain place and on a certain date. It is a separable accident of an individual that he resides at a certain place and is of a certain age.

§ 323. There are some remarks which it may be well to make about the above five terms before we pass on to investigate the principle upon which the division is based.

§ 324. In the first place, it must of course be borne in mind that genus and species are relative terms. No class in itself can be either a genus or a species; it only becomes so in reference to some other class, as standing to it in the relation of containing or contained.

§ 325. Again, the distinction between genus and difference on the one hand and property on the other is wholly relative to an assumed definition. When we say 'Man is an animal,' 'Man is rational,' 'Man is progressive,' there is nothing in the nature of these statements themselves to tell us that the predicate is genus, difference, or property respectively. It is only by a tacit reference to the accepted definition of man that this becomes evident to us, Similarly, we cannot know beforehand that the fact of a triangle having three sides is its difference, and the fact of its having three angles a property. It is only when we assume the definition of a triangle as a three-sided figure that the fact of its having three angles sinks into a property. Had we chosen to define it, in accordance with its etymological meaning, as a figure with three angles, its three-sidedness would then have been a mere property, instead of being the difference; for these two attributes are so connected together that, whichever is postulated, the other will necessarily follow.

§ 326. Lastly, it must be noticed that we have not really defined the term 'accident,' not having stated what it is, but only what it is not. It has in fact been reserved as a residual head to cover any attribute which is neither a difference nor a property.

§ 327. If the five heads of predicables above given were offered to us as an exhaustive classification of the possible relations in which the predicate can stand to the subject in a proposition, the first thing that would strike us is that they do not cover the case in which the predicate is a singular term. In such a proposition as 'This man is John,' we have neither a predication of genus or species nor of attribute: but merely the identification of one term with another, as applying to the same object. Such criticism as this, however, would be entirely erroneous, since no singular term was regarded as a predicate. A predicable was another name for a universal, the common term being called a predicable in one relation and a universal in another-a predicable, extensively, in so far as it was applicable to several different things, a universal, intensively, in so far as the attributes indicated were implied in several other notions, as the attributes indicated by 'animal' are implied in 'horse,' 'sheep,' 'goat,' &c.

§ 328. It would be less irrelevant to point out how the classification breaks down in relation to the singular term as subject. When, for instance, we say 'Socrates is an animal,' 'Socrates is a man,' there is nothing in the proposition to show us whether the predicate is a genus or a species: for we have not here the relation of class to class, which gives us genus or species according to their relative extension, but the relation of a class to an individual.

§ 329. Again, when we say

(1) Some animals are men,

(2) Some men are black,

what is there to tell us that the predicate is to be regarded in the one case as a species and in the other as an accident of the subject? Nothing plainly but the assumption of a definition already known.

§ 330. But if this assumption be granted, the classification seems to admit of a more or less complete defense by logic.

For, given any subject, we can predicate of it either a class or an attribute.

When the predicate is a class, the term predicated is called a Genus, if the subject itself be a class, or a Species, if it be an individual.

When, on the other hand, the predicate is an attribute, the attribute predicated may be either the very attribute which distinguishes the subject from other members of the same class, in which case it is called the Difference, or it may be some attribute connected with the definition, i.e. Property, or not connected with it, i.e. Accident.

§ 331. These results may be exhibited in the following scheme—

```
                      Predicate
       _____|_____
      | |
      Class Attribute
    _____|_____   _____|_____
    | | | |
```

```
(Subject a (Subject a (The (Not the
common singular distinguishing distinguishing
term) term) Attribute) attribute)
Genus    Species    Difference
                |_____
                | |
         (Connected (Not connected
          with the   with the
          definition) definition)
          Property   Accident
```

§ 332. The distinction which underlies this division between predicating a class and predicating an attribute (in quid or in quale) is a perfectly intelligible one, corresponding as it does to the grammatical distinction between the predicate being a noun substantive or a noun adjective. Nevertheless it is a somewhat arbitrary one, since, even when the predicate is a class-name, what we are concerned to convey to the mind, is the fact that the subject possesses the attributes which are connoted by that class-name. We have not here the difference between extensive and intensive predication, since, as we have already seen (§ 264), that is not a difference between one proposition and another, but a distinction in our mode of interpreting any and every proposition. Whatever proposition we like to take may be read either in extension or in intension, according as we fix our minds on the fact of inclusion in a class or the fact of the possession of attributes.

§ 333. It will be seen that the term 'species,' as it appears in the scheme, has a wholly different meaning from the current acceptation in which it was defined above. Species, in its now accepted meaning, signifies the relation of a smaller class to a larger one: as it was originally intended in the heads of predicables it signifies a class in reference to individuals.

§ 334. Another point which requires to be noticed with regard to this five-fold list of heads of predicables, if its object be to classify the relations of a predicate to a subject, is that it takes no account of those forms of predication in which class and attribute are combined. Under which of the five heads would the predicates in the following propositions fall?

(1) Man is a rational animal.

(2) Man is a featherless biped.

In the one case we have a combination of genus and difference; in the other we have a genus combined with an accident.

§ 335. The list of heads of predicables which we have been discussing is not derived from Aristotle, but from the 'Introduction' of Porphyry, a Greek commentator who lived more than six centuries later.

Aristotle's Heads of Predicables.

§ 336. Aristotle himself, by adopting a different basis of division, has allowed room in his classification for the mixed forms of predication above alluded to. His list contains only four heads, namely,

Genus ([Greek: génos])

Definition ([Greek: òrismós])

Proprium ([Greek: îdion])

Accident ([Greek: sumbebekós])

§ 337. Genus here is not distinguished from difference. Whether we say 'Man is an animal' or 'Man is rational,' we are equally understood to be predicating a genus.

§ 338. There is no account taken of species, which, when predicated, resolves itself either into genus or accident. When predicated of an individual, it is regarded as a genus, e.g. 'Socrates is a man'; when predicated of a class, it is regarded as an accident, e.g. 'Some animals are men.'

§ 339. Aristotle's classification may easily be seen to be exhaustive. For every predicate must either be coextensive with its subject or not, i.e. predicable of the same things. And if the two terms coincide in extension, the predicate must either coincide also in intension with the subject or not.

A predicate which coincides both in extension and intension with its subject is exactly what is meant by a definition. One which coincides in extension without coinciding in intension, that is, which applies to the same things without expressing the whole meaning, of the subject, is what is known as a Proprium or Peculiar Property.

If, on the other hand, the two terms are not co-extensive, the predicate must either partially coincide in intension with the subject or not. [Footnote: The case could not arise of a predicate which was entirely coincided in intension with a subject with which it was not co-extensive. For, if the extension of the predicate were greater than that of the subject, its intension would be less, and if less, greater, in accordance with the law of inverse variation of the two quantities (§ 166).] This is equivalent to saying that it must either state part of the definition of the subject or not. Now the definition is made up of genus and difference, either of which may form the predicate: but as the two are indistinguishable in relation to a single subject, they are lumped together for the present purpose under the one head, genus. When the predicate, not being co-extensive, is not even partially co-intensive with its subject, it is called an Accident.

§ 340. Proprium, it will be seen, differs from property. A proprium is an attribute which is possessed by all the members of a class, and by them alone, e.g. 'Men are the only religious animals.'

§ 341. Under the head of definition must be included all propositions in which the predicate is a mere synonym of the subject, e.g. 'Naso is Ovid,' 'A Hebrew is a Jew,' 'The skipper is the captain.' In such propositions the predicate coincides in extension with the subject, and may be considered to coincide in intension where the intension of both subject and predicate is at zero, as in the case of two proper names.

§ 342. Designations and descriptions will fall under the head of 'proprium' or peculiar property, e.g. 'Lord Salisbury is the present prime minister of England,' 'Man is a mammal with hands and without hair.' For here, while the terms are coincident in extension, they are far from being so in intension.

§ 343. The term 'genus' must be understood to include not only genus in the accepted sense, but difference and generic property as well.

§ 344. These results may be exhibited in the following scheme—

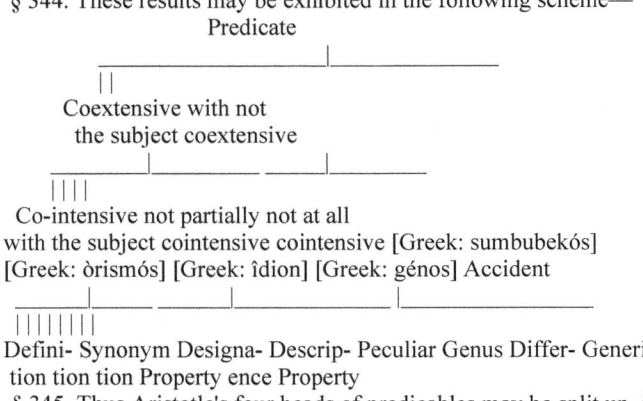

Predicate

Coextensive with / not the subject coextensive

Co-intensive / not partially / not at all
with the subject cointensive / cointensive [Greek: sumbubekós]
[Greek: òrismós] [Greek: îdion] [Greek: génos] Accident

Defini- Synonym Designa- Descrip- Peculiar Genus Differ- Generic
tion tion tion Property ence Property

§ 345. Thus Aristotle's four heads of predicables may be split up, if we please, into nine—

1. Definition \
      > [Greek: òrismós].
2. Synonym /
  3. Designation \
       |
4. Description > [Greek: îdion].
       |
5. Peculiar Property/
  6. Genus \
       |
7. Difference > [Greek: génos].
       |
8. Generic Property/
  9. Accident—[Greek: sumbebekós].

§ 346. We now pass on to the two subjects of Definition and Division, the discussion of which will complete our treatment of the second part of logic. Definition and division correspond respectively to the two kinds of quantity possessed by terms.

Definition is unfolding the quantity of a term in intension.

Division is unfolding the quantity of a term in extension.

CHAPTER VII.

Of Definition.

§ 347. To define a term is to unfold its intension, i.e. to explain its meaning.

§ 348. From this it follows that any term which possesses no intension cannot be defined.

§ 349. Hence proper names do not admit of definition, except just in so far as they do possess some slight degree of intension: Thus we can define the term 'John' only so far as to say that 'John' is the name of a male person. This is said with regard to the original intension of proper names; their acquired intension will be considered later.

§ 350. Again, since definition is unfolding the intension of a term, it follows that those terms will not admit of being defined whose intension is already so simple that it cannot be unfolded further. Of this nature are names of simple attributes, such as greenness, sweetness, pleasure, existence. We know what these things are, but we cannot define them. To a man who has never enjoyed sight, no language can convey an idea of the greenness of the grass or the blueness of the sky; and if a person were unaware of the meaning of the term 'sweetness,' no form of words could convey to him an idea of it. We might put a lump of sugar into his mouth, but that would not be a logical definition.

§ 351. Thus we see that, for a thing to admit of definition, the idea of it must be complex. Simple ideas baffle definition, but at the same time do not require it. In defining we lay out the simpler ideas which are combined in our notion of something, and so explain that complex notion. We have defined 'triangle,' when we analyse it into 'figure' and 'contained by three lines.' Similarly we have defined 'substance' when we analyse it into 'thing' and 'which can be conceived to exist by itself.'

§ 352. But when we get to 'thing' we have reached a limit. The Summum Genus, or highest class under which all things fall, cannot be defined any more than a simple attribute; and for the very good reason that it connotes nothing but pure being, which is the simplest of all attributes. To say that a thing is an 'object of thought' is not really to define it, but to explain its etymology, and to reclaim a philosophical term from its abuse by popular language, in which it is limited to the concrete and the lifeless. Again, to define it negatively and to say that a thing is 'that which is not nothing' does not carry us any further than we were before. The law of contradiction warrants us in saying as much as that.

§ 353. Definition is confined to subject-terms, and does not properly extend to attributives. For definition is of things through names, and an attributive out of predication is not the name of anything. The attributive is defined, so far as it can be, through the corresponding abstract term.

§ 354. Common terms, other than attributives, ought always to admit of definition. For things are distributed by the mind into classes owing to their possessing certain attributes in common, and the definition of the class-name can be effected by detailing these attributes, or at least a sufficient number of them.

§ 355. It is different with singular terms. Singular terms, when abstract, admit of definition, in so far as they are not names of attributes so simple as to evade analysis. When singular terms are concrete, we have to distinguish between the two cases of proper names and designations. Designations are connotative singular terms. They are formed by limiting a common term to the 'case in hand.' Whatever definition therefore fits the common term will fit also the designation which is formed from it, if we add the attributes implied by the limitations. Thus whatever definition fits the common term 'prime minister' will fit also the singular term 'the present prime minister of England' by the addition to it of the attributes of place and time which are indicated by the expression. Such terms as this have a definite amount of intension, which can therefore be seized upon and expounded by a definition.

§ 356. But proper names, having no original intension of their own, cannot be defined at all; whereas, if we look upon them from the point of view of their acquired intension, they defy definition by reason of the very complexity of their meaning. We cannot say exactly what 'John' and 'Mary' mean, because those names, to us who know the particular persons denoted by them, suggest all the most trifling accidents of the individual as well as the essential attributes of the genus.

§ 357. Definition serves the practical purpose of enabling us mentally to distinguish, or, as the name implies, 'mark off' the thing defined from all other things whatsoever. This may seem at first an endless task, but there is a short cut by which the goal may be reached. For, if we distinguish the thing in hand from the things which it is most like, we shall, 'a fortiori,' have distinguished it from things to which it bears a less resemblance.

§ 358. Hence the first thing to do in seeking for a definition is to fix upon the class into which the thing to be defined most naturally falls, and then to distinguish the thing in question from the other members of that class. If we were asked to define a triangle, we would not begin by distinguishing it from a hawser, but from a square and other figures with which it is more possible to confound it. The class into which a thing falls is called its Genus, and the attribute or attributes which distinguish it from other members of that class are called its Difference.

§ 359. If definition thus consists in referring a thing to a class, we see a further reason why the summum genus of all things cannot be defined.

§ 360. We have said that definition is useful in enabling us to distinguish things from one another in our minds: but this must not be regarded as the direct object of the process. For this object may be accomplished without giving a definition at all, by means of what is called a Description. By a description is meant an enumeration of accidents with or without the mention of some class-name. It is as applicable to proper names as to common terms. When we say 'John Smith lives next door on the right-hand side and passes by to his office every morning at nine o'clock,' we have, in all probability, effectually distinguished John Smith from other people: but living next, &c., cannot be part of the intension of John Smith, since John Smith may change his residence or abandon his occupation without ceasing to be called by his name. Indirectly then definition serves the purpose of distinguishing things in the mind, but its direct object is to unfold the intension of terms, and so impart precision to our thoughts by setting plainly before us the meaning of the words we are using.

§ 361. But when we say that definition is unfolding the intension of terms, it must not be imagined that we are bound in defining to unfold completely the intension of terms. This would be a tedious, and often an endless, task. A term may mean, or convey to the mind, a good many more attributes than those which are stated in its definition. There is no limit indeed to the meaning which a term may legitimately convey, except the common attributes of the things denoted by it. Who shall say, for instance, that a triangle means a figure with three sides, and does not mean a figure with three angles, or the surface of the perpendicular bisection of a cone? Or again, that man means a rational, and does not mean a speaking, a religious, or an aesthetic animal, or a biped with two eyes, a nose, and a mouth? The only attributes of which it can safely be asserted that they can form no part of the intension of a term are those which are not common to all the things to which the name applies. Thus a particular complexion, colour, height, creed, nationality cannot form any part of the intension of the term 'man.' But among the attributes common to a class we cannot distinguish between essential and unessential, except by the aid of definition itself. Formal logic cannot recognise any order of priority between the attributes common to all the members of a class, such as to necessitate our recognising some as genera and differentiae and relegating others to the place of properties or inseparable accidents.

§ 362. The art of giving a good definition is to seize upon the salient characteristics of the thing defined and those wherefrom the largest number of other attributes can be deduced as consequences. To do this well requires a special knowledge of the thing in question, and is not the province of the formal logician.

§ 363. We have seen already, in treating of the Heads of Predicables (§ 325), that the difference between genus and difference on the one hand and property on the other is wholly relative to some assumed definition. Now definitions are always to a certain extent arbitrary, and will vary with the point of view from which we consider the thing required to be defined. Thus 'man' is usually contrasted with 'brute,' and from this point of view it is held a sufficient definition of him to say that he is 'a rational animal,' But a theologian might be more anxious to contrast man with supposed incorporeal intelligences, and from this point of view man would be defined as an 'embodied spirit.'

§ 364. In the two definitions just given it will be noticed that we have really employed exactly the same attributes, only their place as genus and difference has been reversed. It is man's rational, or spiritual, nature which distinguishes him from the brutes: but this is just what he is supposed to have in common with incorporeal intelligences, from whom he is differentiated by his animal nature.

[Illustration]

This illustration is sufficient to show us that, while there is no absolute definition of anything, in the sense of a fixed genus and difference, there may at the same time be certain attributes which permanently distinguish the members of a given class from those of all other classes.

§ 365. The above remarks will have made it clear that the intension of a term is often much too wide to be conveyed by any definition; and that what a definition generally does is to select certain attributes from the whole intension, which are regarded as being more typical of the thing than the remainder. No definition can be expected to exhaust the whole intension of a term, and there will always be room for varying definitions of the same thing, according to the different points of view from which it is approached.

§ 366. Names of attributes lend themselves to definition far more easily than names of substances. The reason of this is that names of attributes are primarily intensive in force, whereas substances are known to us in extension before they become known to us in intension. There is no difficulty in defining a term like 'triangle' or 'monarchy,' because these terms were expressly invented to cover certain attributes; but the case is different with such terms as 'dog,' 'tree,' 'plant,' 'metal,' and other names of concrete things. We none of us have any difficulty in recognising a dog or tree, when we see them, or in distinguishing them from other animals or plants respectively. We are therefore led to imagine that we know the meaning of these terms. It is not until we are called upon for a definition that we discover how superficial our knowledge really is of the common attributes possessed by the things which these names denote.

§ 367. It might be imagined that a common name would never be given to things except in virtue of our knowledge of their common attributes. But as a matter of fact, the common name was first given from a confused notion of resemblance, and we had afterwards to detect the common attributes, when sometimes the name had been so extended from one thing to another like it, that there were hardly any definite attributes possessed in common by the earlier and later members of the class.

§ 368. This is especially the case where the meaning of terms has been extended by analogy, e.g. head, foot, arm, post, pole, pipe, &c.

§ 369. But in the progress of thought we come to form terms in which the intensive capacity is everything. Of this kind notably are mathematical conceptions. Terms of this kind, as we said before, lend themselves readily to definition.

§ 370. We may lay down then roughly that words are easy or difficult of definition according as their intensive or extensive capacity predominates.

§ 371. There is a marked distinction to be observed between the classes made by the mind of man and the classes made by nature, which are known as 'real kinds.' In the former there is generally little or nothing in common except the particular attribute which is selected as the ground of classification, as in the case of red and white things, which are alike only in their redness or whiteness; or else their attributes are all necessarily connected, as in the case of circle, square and triangle. But the members of nature's classes agree in innumerable attributes which have no discoverable connection with one another, and which must therefore, provisionally at least, be regarded as standing in the relation of inseparable accidents to any particular attributes which we may select for the purposes of definition. There is no assignable reason why a rational animal should have hair on its head or a nose on its face, and yet man, as a matter of fact, has both; and generally the particular bodily configuration of man can only be regarded as an inseparable accident of his nature as a rational animal.

§ 372. 'Real kinds' belong to the class of words mentioned above in which the extension predominates over the intension. We know well enough the things denoted by them, while most of us have only a dim idea of the points of resemblance between these things. Nature's classes moreover shade off into one another by such imperceptible degrees that it is often impossible to fix the boundary line between one class and another. A still greater source of perplexity in dealing with real kinds is that it is sometimes almost impossible to fix upon any attribute which is common to every individual member of the class without exception. All that we can do in such cases is to lay down a type of the class in its perfect form, and judge of individual instances by the degree of their approximation to it. Again, real kinds being known to us primarily in extension, the intension which we attach to the names is liable to be affected by the advance of knowledge. In dealing therefore with such terms we must be content with provisional definitions, which adequately express our knowledge of the things denoted by them, at the time, though a further study of their attributes may induce us subsequently to alter the definition. Thus the old definition of animal as a sentient organism has been rendered inadequate by the discovery that so many of the phenomena of sensation can be exhibited by plants,

§ 373. But terms in which intension is the predominant idea are more capable of being defined once for all. Aristotle's definitions of 'wealth' and 'monarchy' are as applicable now as in his own day, and no subsequent discoveries of the properties of figures will render Euclid's definitions unavailable.

§ 374. We may distinguish therefore between two kinds of definition, namely,

(1) Final.

(2) Provisional.

§ 375. A distinction is also observed between Real and Nominal Definitions. Both of these explain the meaning of a term: but a real definition further assumes the actual existence of the thing defined. Thus the explanation of the term 'Centaur' would be a nominal, that of 'horse' a real definition.

It is useless to assert, as is often done, that a nominal definition explains the meaning of a term and a real definition the nature of a thing; for, as we have seen already, the meaning of a term is whatever we know of the nature of a thing.

§ 376. It now remains to lay down certain rules for correct definition.

§ 377. The first rule that is commonly given is that a definition should state the essential attributes of the thing defined. But this amounts merely to saying that a definition should be a definition; since it is only by the aid of definition that we can distinguish between essential and non-essential among the common attributes exhibited by a class of things. The rule however may be retained as a material test of the soundness of a definition, in the sense that he who seeks to define anything should fix upon its most important attributes. To define man as a mammiferous animal having two hands, or as a featherless biped, we feel to be absurd and incongruous, since there is no reference to the most salient characteristic of man, namely, his rationality. Nevertheless we cannot quarrel with these definitions on formal, but only on material grounds. Again, if anyone chose to define logic as the art of thinking, all we could say is that we differ from him in opinion, as we think logic is more properly to be regarded as the science of the laws of thought. But here also it is on material grounds that we dissent from the definition.

§ 378. Confining ourselves therefore to the sphere with which we are properly concerned, we lay down the following

Rules for Definition.

(1) A definition must be co-extensive with the term defined.

(2) A definition must not state attributes which imply one another.

   (3) A definition must not contain the name defined, either directly or by implication.

(4) A definition must be clearer than the term defined.

(5) A definition must not be negative, if it can be affirmative.

Briefly, a definition must be adequate (1), terse (2), clear (4); and must not be tautologous (3), or, if it can be avoided, negative (5).

§ 379. It is worth while to notice a slight ambiguity in the term 'definition' itself. Sometimes it is applied to the whole proposition which expounds the meaning of the term; at other times it is confined to the predicate of this proposition. Thus in stating the first four rules we have used the term in the latter sense, and in stating the fifth in the former.

§ 380. We will now illustrate the force of the above rules by giving examples of their violation.

Rule 1. Violations. A triangle is a figure with three equal sides.

A square is a four-sided figure having all its sides equal.

In the first instance the definition is less extensive than the term defined, since it applies only to equilateral triangles. This fault may be amended by decreasing the intension, which we do by eliminating the reference to the equality of the sides.

In the second instance the definition is more extensive than the term defined. We must accordingly increase the intension by adding a new attribute 'and all its angles right angles.'

Rule 2. Violation. A triangle is a figure with three sides and three angles.

One of the chief merits of a definition is to be terse, and this definition is redundant, since what has three sides cannot but have three angles.

Rule 3. Violations. A citizen is a person both of whose parents were citizens.
Man is a human being.

Rule 4. Violations. A net is a reticulated fabric, decussated at regular intervals.

Life is the definite combination of heterogeneous changes, both simultaneous and successive, in correspondence with external co-existences and sequences.

Rule 5. Violations. A mineral is that which is neither animal nor vegetable.
Virtue is the absence of vice.

§ 381. The object of definition being to explain what a thing is, this object is evidently defeated, if we confine ourselves to saying what it is not. But sometimes this is impossible to be avoided. For there are many terms which, though positive in form, are privative in force. These terms serve as a kind of residual heads under which to throw everything within a given sphere, which does not exhibit certain positive attributes. Of this unavoidably negative nature was the definition which we give of 'accident,' which amounted merely to saying that it was any attribute which was neither a difference nor a property.

§ 382. The violation of Rule 3, which guards against defining a thing by itself, is technically known as 'circulus in definiendo,' or defining in a circle. This rule is often apparently violated, without being really so. Thus Euclid defines an acute-angled triangle as one which has three acute angles. This seems a glaring violation of the rule, but is perfectly correct in its context; for it has already been explained what is meant by the terms 'triangle' and 'acute angle,' and all that is now required is to distinguish the acute-angled triangle from its cognate species. He might have said that an acute-angled triangle is one which has neither a right angle nor an obtuse angle: but rightly preferred to throw the same statement into a positive form.

§ 383. The violation of Rule 4 is known as 'ignotum per ignotius' or 'per aeque ignotum.' This rule also may seemingly be violated when it is not really so. For a definition may be correct enough from a special point of view, which, apart from that particular context, would appear ridiculous. From the point of view of conic sections, it is correct enough to define a triangle as that section of a cone which is formed by a plane passing through the vertex perpendicularly to the base, but this could not be expected to make things clearer to a person who was inquiring for the first time into the meaning of the word triangle. But a real violation of the fourth rule may arise, not only from obscurity, but from the employment of ambiguous language or metaphor. To say that 'temperance is a harmony of the soul' or that 'bread is the staff of life,' throws no real light upon the nature of the definiend.

§ 384. The material correctness of a definition is, as we have already seen, a matter extraneous to formal logic. An acquaintance with the attributes which terms imply involves material knowledge quite as much as an acquaintance with the things they apply to; knowledge of the intension and of the extension of terms is alike acquired by experience. No names are such that their meaning is rendered evident by the very constitution of our mental faculties; yet nothing short of this would suffice to bring the material content of definition within the province of formal logic.

CHAPTER VIII.

Of Division.

§ 385. To divide a term is to unfold its extension, that is, to set forth the things of which it is a name.

§ 386. But as in definition we need not completely unfold the intension of a term, so in division we must not completely unfold its extension.

§ 387. Completely to unfold the extension of a term would involve stating all the individual objects to which the name applies, a thing which would be impossible in the case of most common terms. When it is done, it is called Enumeration. To reckon up all the months of the year from January to December would be an enumeration, and not a division, of the term 'month.'

§ 388. Logical division always stops short at classes. It may be defined as the statement of the various classes of things that can be called by a common name. Technically we may say that it consists in breaking up a genus into its component species.

§ 389. Since division thus starts with a class and ends with classes, it is clear that it is only common terms which admit of division, and also that the members of the division must themselves be common terms.

§ 390. An 'individual' is so called as not admitting of logical division. We may divide the term 'cow' into classes, as Jersey, Devonshire, &c., to which the name 'cow' will still be applicable, but the parts of an individual cow are no longer called by the name of the whole, but are known as beefsteaks, briskets, &c.

§ 391. In dividing a term the first requisite is to fix upon some point wherein certain members of the class differ from others. The point thus selected is called the Fundamentum Divisionis or Basis of the Division.

§ 392. The basis of the division will of course differ according to the purpose in hand, and the same term will admit of being divided on a number of different principles. Thus we may divide the term 'man,' on the basis of colour, into white, black, brown, red, and yellow; or, on the basis of locality, into Europeans, Asiatics, Africans, Americans, Australians, New Zealanders, and Polynesians; or again, on a very different principle, into men of nervous, sanguine, bilious, lymphatic and mixed temperaments.

§ 393. The term required to be divided is known as the Totum Divisum or Divided Whole. It might also be called the Dividend.

§ 394. The classes into which the dividend is split up are called the Membra Dividentia, or Dividing Members.

§ 395. Only two rules need be given for division—

(1) The division must be conducted on a single basis.

(2) The dividing members must be coextensive with the divided whole.

§ 396. More briefly, we may put the same thing thus—There must be no cross-division (1) and the division must be exhaustive (2).

§ 397. The rule, which is commonly given, that each dividing member must be a common term, is already provided for under our definition of the process.

§ 398. The rule that the dividend must be predicable of each of the dividing members is contained in our second rule; since, if there were any term of which the dividend were not predicable, it would be impossible for the dividing members to be exactly coextensive with it. It would not do, for instance, to introduce mules and donkeys into a division of the term horse.

§ 399. Another rule, which is sometimes given, namely, that the constituent species must exclude one another, is a consequence of our first; for, if the division be conducted on a single principle, the constituent species must exclude one another. The converse, however, does not hold true. We may have a division consisting of mutually exclusive members, which yet involves a mixture of different bases, e.g. if we were to divide triangle into scalene, isosceles and equiangular. This happens because two distinct attributes may be found in invariable conjunction.

§ 400. There is no better test, however, of the soundness of a division than to try whether the species overlap, that is to say, whether there are any individuals that would fall into two or more of the classes. When this is found to be the case, we may be sure that we have mixed two or more different fundamenta divisionis. If man, for instance, were to be divided into European, American, Aryan, and Semitic, the species would overlap; for both Europe and America contain inhabitants of Aryan and Semitic origin. We have here members of a division based on locality mixed up with members of another division, which is based on race as indicated by language.

§ 401. The classes which are arrived at by an act of division may themselves be divided into smaller classes. This further process is called Subdivision.

§ 402. Let it be noticed that Rule 1 applies only to a single act of division. The moment that we begin to subdivide we not only may, but must, adopt a new basis of division; since the old one has, 'ex hypothesi,' been exhausted. Thus, having divided men according to the colour of their skins, if we wish to subdivide any of the classes, we must look out for some fresh attribute wherein some men of the same complexion differ from others, e.g. we might divide black men into woolly-haired blacks, such as the Negroes, and straight-haired blacks, like the natives of Australia.

§ 403. We will now take an instance of division and subdivision, with a view to illustrating some of the technical terms which are used in connection with the process. Keeping closely to our proper subject, we will select as an instance a division of the products of thought, which it is the province of logic to investigate.

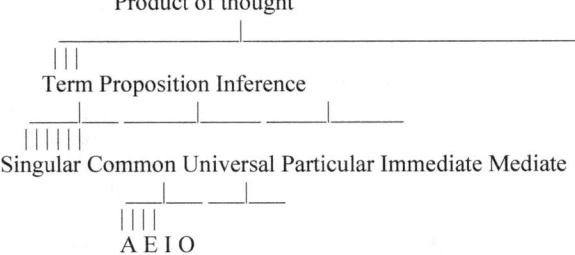

Product of thought

Term Proposition Inference

Singular Common Universal Particular Immediate Mediate

A E I O

Here we have first a threefold division of the products of thought based on their comparative complexity. The first two of these, namely, the term and the proposition, are then subdivided on the basis of their respective quantities. In the case of inference the basis of the division is again the degree of complexity. The subdivision of the proposition is carried a step further than that of the others. Having exhausted our old basis of quantity, we take a new attribute, namely, quality, on which to found the next step of subdivision.

§ 404. Now in such a scheme of division and subdivision as the foregoing, the highest class taken is known as the Summum Genus. Thus the summum genus is the same thing as the divided whole, viewed in a different relation. The term which is called the divided whole with reference to a single act of division, is called the summum genus whenever subdivision has taken place.

§ 405. The classes at which the division stops, that is, any which are not subdivided, are known as the Infimae Species.

§ 406. All classes intermediate between the summum genus and the infimae species are called Subaltern Genera or Subaltern Species, according to the way they are looked at, being genera in relation to the classes below them and species in relation to the classes above them.

§ 407. Any classes which fall immediately under the same genus are called Cognate Species, e.g. singular and common terms are cognate species of term.

§ 408. The classes under which any lower class successively falls are called Cognate Genera. The relation of cognate species to one another is like that of children of the same parents, whereas cognate genera resemble a line of ancestry.

§ 409. The Specific Difference of anything is the attribute or attributes which distinguish it from its cognate species. Thus the specific difference of a universal proposition is that the predicate is known to apply to the whole of the subject. A specific difference is said to constitute the species.

§ 410. The specific difference of a higher class becomes a Generic Difference with respect to the class below it. A generic difference then may be said to be the distinguishing attribute of the whole class to which a given species belongs. The generic difference is common to species that are cognate to one another, whereas the specific difference is peculiar to each. It is the generic difference of an A proposition that it is universal, the specific difference that it is affirmative.

§ 411. The same distinction is observed between the specific and generic properties of a thing. A Specific Property is an attribute which flows from the difference of a thing itself; a Generic Property is an attribute which flows from the difference of the genus to which the thing belongs. It is a specific property of an E proposition that its predicate is distributed, a generic property that its contrary cannot be true along with it (§ 465); for this last characteristic flows from the nature of the universal proposition generally.

§ 412. It now remains to say a few words as to the place in logic of the process of division. Since the attributes in which members of the same class differ from one another cannot possibly be indicated by their common name, they must be sought for by the aid of experience; or, to put the same thing in other words, since all the infimae species are alike contained under the summum genus, their distinctive attributes can be no more than separable accidents when viewed in relation to the summum genus. Hence division, being always founded on the possession or non-possession of accidental attributes, seems to lie wholly outside the sphere of formal logic. This however is not quite the case, for, in virtue of the Law of Excluded Middle, there is always open to us, independently of experience, a hypothetical division by dichotomy. By dichotomy is meant a division into two classes by a pair of contradictory terms, e.g. a division of the class, man, into white and not-white. Now we cannot know, independently of experience, that any members of the class, man, possess whiteness; but we may be quite sure, independently of all experience, that men are either white or not. Hence division by dichotomy comes strictly within the province of formal logic. Only it must be noticed that both sides of the division must be hypothetical. For experience alone can tell us, on the one hand, that there are any men that are white, and on the other, that there are any but white men.

§ 413. What we call a division on a single basis is in reality the compressed result of a scheme of division and subdivision by dichotomy, in which a fresh principle has been introduced at every step. Thus when we divide men, on the basis of colour, into white, black, brown, red and yellow, we may be held to have first divided men into white and not-white, and then to have subdivided the men that are not-white into black and not-black, and so on. From the strictly formal point of view this division can only be represented as follows—

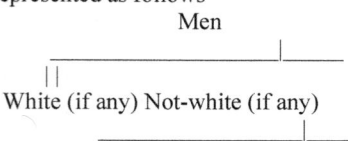

Men

White (if any) Not-white (if any)

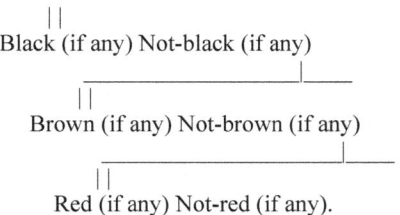

Black (if any) Not-black (if any)

Brown (if any) Not-brown (if any)

Red (if any) Not-red (if any).

§ 414. Formal correctness requires that the last term in such a series should be negative. We have here to keep the term 'not-red' open, to cover any blue or green men that might turn up. It is only experience that enables us to substitute the positive term 'yellow' for 'not-red,' since we know as a matter of fact that there are no men but those of the five colours given in the original division.

§ 415. Any correct logical division always admits of being arrived at by the longer process of division and subdivision by dichotomy. For instance, the term quadrilateral, or four-sided rectilinear figure, is correctly divided into square, oblong, rhombus, rhomboid and trapezium. The steps of which this division consists are as follows—

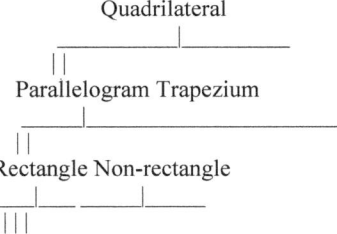

Quadrilateral

Parallelogram Trapezium

Rectangle Non-rectangle

Square Oblong Rhombus Rhomboid.

§ 416. In reckoning up the infimae species in such a scheme, we must of course be careful not to include any class which has been already subdivided; but no harm is done by mixing an undivided class, like trapezium, with the subdivisions of its cognate species.

§ 417. The two processes of definition and division are intimately connected with one another. Every definition suggests a division by dichotomy, and every division supplies us at once with a complete definition of all its members.

§ 418. Definition itself, so far as concerns its content, is, as we have already seen, extraneous to formal logic: but when once we have elicited a genus and difference out of the material elements of thought, we are enabled, without any further appeal to experience, to base thereon a division by dichotomy. Thus when man has been defined as a rational animal, we have at once suggested to us a division of animal into rational and irrational.

§ 419. Again, the addition of the attributes, rational and irrational respectively, to the common genus, animal, ipso facto supplies us with definitions of the species, man and brute. Similarly, when we subdivided rectangle into square and oblong on the basis of the equality or inequality of the adjacent sides, we were by so doing supplied with a definition both of square and oblong—'A square is a rectangle having all its sides equal,' and 'An oblong is a rectangle which has only its opposite sides equal.'

§ 420. The definition of a square just given amounts to the same thing as Euclid's definition, but it complies with a rule which has value as a matter of method, namely, that the definition should state the Proximate Genus of the thing defined.

§ 421. Since definition and division are concerned with the intension and extension of terms, they are commonly treated of under the first part of logic: but as the treatment of the subject implies a familiarity with the Heads of Predicables, which in their turn imply the proposition, it seems more desirable to deal with them under the second.

§ 422. We have already had occasion to distinguish division from Enumeration. The latter is the statement of the individual things to which a name applies. In enumeration, as in division, the wider term is predicable of each of the narrower ones.

§ 423. Partition is the mapping out of a physical whole into its component parts, as when we say that a tree consists of roots, stem, and branches. In a partition the name of the whole is not predicable of each of the parts.

§ 424. Distinction is the separation from one another of the various meanings of an equivocal term. The term distinguished is predicable indeed of each of the members, but of each in a different sense. An equivocal term is in fact not one but several terms, as would quickly appear, if we were to use definitions in place of names.

§ 425. We have seen that a logical whole is a genus viewed in relation to its underlying species. From this must be distinguished a metaphysical whole, which is a substance viewed in relation to its attributes, or a class regarded in the same way. Logically, man is a part of the class, animal; metaphysically, animal is contained in man. Thus a logical whole is a whole in extension, while a metaphysical whole is a whole in intension. From the former point of view species is contained in genus; from the latter genus is contained in species.

PART III.—OF INFERENCES.

CHAPTER I.

Of Inferences in General.

§ 426. To infer is to arrive at some truth, not by direct experience, but as a consequence of some truth or truths already known. If we see a charred circle on the grass, we infer that somebody has been lighting a fire there, though we have not seen anyone do it. This conclusion is arrived at in consequence of our previous experience of the effects of fire.

§ 427. The term Inference is used both for a process and for a product of thought.

As a process inference may be defined as the passage of the mind from one or more propositions to another.

As a product of thought inference may be loosely declared to be the result of comparing propositions.

§ 428. Every inference consists of two parts—

(1) the truth or truths already known;

(2) the truth which we arrive at therefrom.

The former is called the Antecedent, the latter the Consequent. But this use of the terms 'antecedent' and 'consequent' must be carefully distinguished from the use to which they were put previously, to denote the two parts of a complex proposition.

§ 429. Strictly speaking, the term inference, as applied to a product of thought, includes both the antecedent and consequent: but it is often used for the consequent to the exclusion of the antecedent. Thus, when we have stated our premises, we say quite naturally, 'And the inference I draw is so and so.'

§ 430. Inferences are either Inductive or Deductive. In induction we proceed from the less to the more general; in deduction from the more to the less general, or, at all events, to a truth of not greater generality than the one from which we started. In the former we work up to general principles; in the latter we work down from them, and elicit the particulars which they contain.

§ 431. Hence induction is a real process from the known to the unknown, whereas deduction is no more than the application of previously existing knowledge; or, to put the same thing more technically, in an inductive inference the consequent is not contained in the antecedent, in a deductive inference it is.

§ 432. When, after observing that gold, silver, lead, and other metals, are capable of being reduced to a liquid state by the application of heat, the mind leaps to the conclusion that the same will hold true of some other metal, as platinum, or of all metals, we have then an inductive inference, in which the conclusion, or consequent, is a new proposition, which was not contained in those that went before. We are led to this conclusion, not by reason, but by an instinct which teaches us to expect like results, under like circumstances. Experience can tell us only of the past: but we allow it to affect our notions of the future through a blind belief that 'the thing that hath been, it is that which shall be; and that which is done is that which shall be done; and there is no new thing under the sun.' Take away this conviction, and the bridge is cut which connects the known with the unknown, the past with the future. The commonest acts of daily life would fail to be performed, were it not for this assumption, which is itself no product of the reason. Thus man's intellect, like his faculties generally, rests upon a basis of instinct. He walks by faith, not by sight.

§ 433. It is a mistake to talk of inductive reasoning, as though it were a distinct species from deductive. The fact is that inductive inferences are either wholly instinctive, and so unsusceptible of logical vindication, or else they may be exhibited under the form of deductive inferences. We cannot be justified in inferring that platinum will be melted by heat, except where we have equal reason for asserting the same thing of copper or any other metal. In fact we are justified in drawing an individual inference only when we can lay down the universal proposition, 'Every metal can be melted by heat.' But the moment this universal proposition is stated, the truth of the proposition in the individual instance flows from it by way of deductive inference. Take away the universal, and we have no logical warrant for arguing from one individual case to another. We do so, as was said before, only in virtue of that vague instinct which leads us to anticipate like results from like appearances.

§ 434. Inductive inferences are wholly extraneous to the science of formal logic, which deals only with formal, or necessary, inferences, that is to say with deductive inferences, whether immediate or mediate. These are called formal, because the truth of the consequent is apparent from the mere form of the antecedent, whatever be the nature of the matter, that is, whatever be the terms employed in the proposition or pair of propositions which constitutes the antecedent. In deductive inference we never do more than vary the form of the truth from which we started. When from the proposition 'Brutus was the founder of the Roman Republic,' we elicit the consequence 'The founder of the Roman Republic was Brutus,' we certainly have nothing more in the consequent than was already contained in the antecedent; yet all deductive inferences may be reduced to identities as palpable as this, the only difference being that in more complicated cases the consequent is contained in the antecedent along with a number of other things, whereas in this case the consequent is absolutely all that the antecedent contains.

§ 435. On the other hand, it is of the very essence of induction that there should be a process from the known to the unknown. Widely different as these two operations of the mind are, they are yet both included under the definition which we have given of inference, as the passage of the mind from one or more propositions to another. It is necessary to point this out, because some logicians maintain that all inference must be from the known to the unknown, whereas others confine it to 'the carrying out into the last proposition of what was virtually contained in the antecedent judgements.'

§ 436. Another point of difference that has to be noticed between induction and deduction is that no inductive inference can ever attain more than a high degree of probability, whereas a deductive inference is certain, but its certainty is purely hypothetical.

§ 437. Without touching now on the metaphysical difficulty as to how we pass at all from the known to the unknown, let us grant that there is no fact better attested by experience than this—'That where the circumstances are precisely alike, like results follow.' But then we never can be absolutely sure that the circumstances in any two cases are precisely alike. All the experience of all past ages in favour of the daily rising of the sun is not enough to render us theoretically certain that the sun will rise tomorrow We shall act indeed with a perfect reliance upon the assumption of the coming day-break; but, for all that, the time may arrive when the conditions of the universe shall have changed, and the sun will rise no more.

§ 438. On the other hand a deductive inference has all the certainty that can be imparted to it by the laws of thought, or, in other words, by the structure of our mental faculties; but this certainty is purely hypothetical. We may feel assured that if the premisses are true, the conclusion is true also. But for the truth of our premisses we have to fall back upon induction or upon intuition. It is not the province of deductive logic to discuss the material truth or falsity of the propositions upon which our reasonings are based. This task is left to inductive logic, the aim of which is to establish, if possible, a test of material truth and falsity.

§ 439. Thus while deduction is concerned only with the relative truth or falsity of propositions, induction is concerned with their actual truth or falsity. For this reason deductive logic has been termed the logic of consistency, not of truth.

§ 440. It is not quite accurate to say that in deduction we proceed from the more to the less general, still less to say, as is often said, that we proceed from the universal to the particular. For it may happen that the consequent is of precisely the same amount of generality as the antecedent. This is so, not only in most forms of immediate inference, but also in a syllogism which consists of singular propositions only, e.g.

The tallest man in Oxford is under eight feet.

So and so is the tallest man in Oxford.

.'. So and so is under eight feet.

This form of inference has been named Traduction; but there is no essential difference between its laws and those of deduction.

§ 441. Subjoined is a classification of inferences, which will serve as a map of the country we are now about to explore.

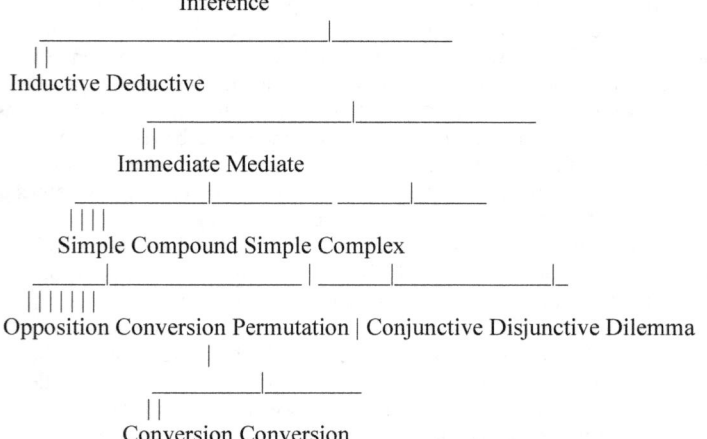

by by
Negation position

CHAPTER II.

Of Deductive Inferences.

§ 442. Deductive inferences are of two kinds—Immediate and Mediate.

§ 443. An immediate inference is so called because it is effected without the intervention of a middle term, which is required in mediate inference.

§ 444. But the distinction between the two might be conveyed with at least equal aptness in this way—

An immediate inference is the comparison of two propositions directly.

A mediate inference is the comparison of two propositions by means of a third.

§ 445. In that sense of the term inference in which it is confined to the consequent, it may be said that—

An immediate inference is one derived from a single proposition.

A mediate inference is one derived from two propositions conjointly.

§ 446. There are never more than two propositions in the antecedent of a deductive inference. Wherever we have a conclusion following from more than two propositions, there will be found to be more than one inference.

§ 447. There are three simple forms of immediate inference, namely Opposition, Conversion and Permutation.

§ 448. Besides these there are certain compound forms, in which permutation is combined with conversion.

CHAPTER III.

Of Opposition.

§ 449. Opposition is an immediate inference grounded on the relation between propositions which have the same terms, but differ in quantity or in quality or in both.

§ 450. In order that there should be any formal opposition between two propositions, it is necessary that their terms should be the same. There can be no opposition between two such propositions as these—

(1) All angels have wings.

(2) No cows are carnivorous.

§ 451. If we are given a pair of terms, say A for subject and B for predicate, and allowed to affix such quantity and quality as we please, we can of course make up the four kinds of proposition recognised by logic, namely,

A. All A is B.

E. No A is B.

I. Some A is B.

O. Some A is not B.

§ 452. Now the problem of opposition is this: Given the truth or falsity of any one of the four propositions A, E, I, O, what can be ascertained with regard to the truth or falsity of the rest, the matter of them being supposed to be the same?

§ 453. The relations to one another of these four propositions are usually exhibited in the following scheme—

A . . . . Contrary . . . . E

. . . .
. . . .
. . . .
. . . .
. . . .
. . . .

Subaltern Contradictory Subaltern

. . . .
. . . .
. . . .
. . . .
. . . .
. . . .

I . . . Sub-contrary . . . O

§ 454. Contrary Opposition is between two universals which differ in quality.

§ 455. Sub-contrary Opposition is between two particulars which differ in quality.

§ 456. Subaltern Opposition is between two propositions which differ only in quantity.

§ 457. Contradictory Opposition is between two propositions which differ both in quantity and in quality.

§ 458. Subaltern Opposition is also known as Subalternation, and of the two propositions involved the universal is called the Subalternant and the particular the Subalternate. Both together are called Subalterns, and similarly in the other forms of opposition the two propositions involved are known respectively as Contraries, Sub-contraries and Contradictories.

§ 459. For the sake of convenience some relations are classed under the head of opposition in which there is, strictly speaking, no opposition at all between the two propositions involved.

§ 460. Between sub-contraries there is an apparent, but not a real opposition, since what is affirmed of one part of a term may often with truth be denied of another. Thus there is no incompatibility between the two statements.

(1) Some islands are inhabited.

(2) Some islands are not inhabited.

§ 461. In the case of subaltern opposition the truth of the universal not only may, but must, be compatible with that of the particular.

§ 462. Immediate Inference by Relation would be a more appropriate name than Opposition; and Relation might then be subdivided into Compatible and Incompatible Relation. By 'compatible' is here meant that there is no conflict between the *truth* of the two propositions. Subaltern and sub-contrary opposition would thus fall under the head of compatible relation; contrary and contradictory relation under that of incompatible relation.

Relation
_____|_____
| |
Compatible Incompatible
_____|_____ _____|_____

| | | |

Subaltern Sub-contrary Contrary Contradictory.

§ 463. It should be noticed that the inference in the case of opposition is from the truth or falsity of one of the opposed propositions to the truth or falsity of the other.

§ 464. We will now lay down the accepted laws of inference with regard to the various kinds of opposition.

§ 465. Contrary propositions may both be false, but cannot both be true. Hence if one be true, the other is false, but not vice versâ.

§ 466. Sub-contrary propositions may both be true, but cannot both be false. Hence if one be false, the other is true, but not vice versâ.

§ 467. In the case of subaltern propositions, if the universal be true, the particular is true; and if the particular be false, the universal is false; but from the truth of the particular or the falsity of the universal no conclusion can be drawn.

§ 468. Contradictory propositions cannot be either true or false together. Hence if one be true, the other is false, and vice versâ.

§ 469. By applying these laws of inference we obtain the following results—

If A be true, E is false, O false, I true.

If A be false, E is unknown, O true, I unknown.

If E be true, O is true, I false, A false.

If E be false, O is unknown, I true, A unknown.

If O be true, I is unknown, A false, E unknown.

If O be false, I is true, A true, E false.

If I be true, A is unknown, E false, O unknown.

If I be false, A is false, E true, O true.

§ 470. It will be seen from the above that we derive more information from deriving a particular than from denying a universal. Should this seem surprising, the paradox will immediately disappear, if we reflect that to deny a universal is merely to assert the contradictory particular, whereas to deny a particular is to assert the contradictory universal. It is no wonder that we should obtain more information from asserting a universal than from asserting a particular.

§ 471. We have laid down above the received doctrine with regard to opposition: but it is necessary to point out a flaw in it.

When we say that of two sub-contrary propositions, if one be false, the other is true, we are not taking the propositions I and O in their now accepted logical meaning as indefinite (§ 254), but rather in their popular sense as 'strict particular' propositions. For if I and O were taken as indefinite propositions, meaning 'some, if not all,' the truth of I would not exclude the possibility of the truth of A, and, similarly, the truth of O would not exclude the possibility of the truth of E. Now A and E may both be false. Therefore I and O, being possibly equivalent to them, may both be false also. In that case the doctrine of contradiction breaks down as well. For I and O may, on this showing, be false, without their contradictories E and A being thereby rendered true. This illustrates the awkwardness, which we have previously had occasion to allude to, which ensures from dividing propositions primarily into universal and particular, instead of first dividing them into definite and indefinite, and particular (§ 256).

§ 472. To be suddenly thrown back upon the strictly particular view of I and O in the special case of opposition, after having been accustomed to regard them as indefinite propositions, is a manifest inconvenience. But the received doctrine of opposition does not even adhere consistently to this view. For if I and O be taken as strictly particular propositions, which exclude the possibility of the universal of the same quality being true along with them, we ought not merely to say that I and O may both be true, but that if one be true the other must also be true. For I being true, A is false, and therefore O is true; and we may argue similarly from the truth of O to the truth of I, through the falsity of E.

Or—to put the Same thing in a less abstract form—since the strictly particular proposition means 'some, but not all,' it follows that the truth of one sub-contrary necessarily carries with it the truth of the other, If we lay down that some islands only are inhabited, it evidently follows, or rather is stated simultaneously, that there are some islands also which are not inhabited. For the strictly particular form of proposition 'Some A only is B' is of the nature of an exclusive proposition, and is really equivalent to two propositions, one affirmative and one negative.

§ 473. It is evident from the above considerations that the doctrine of opposition requires to be amended in one or other of two ways. Either we must face the consequences which follow from regarding I and O as indefinite, and lay down that sub-contraries may both be false, accepting the awkward corollary of the collapse of the doctrine of contradiction; or we must be consistent with ourselves in regarding I and O, for the particular purposes of opposition, as being strictly particular, and lay down that it is always possible to argue from the truth of one sub-contrary to the truth of the other. The latter is undoubtedly the better course, as the admission of I and O as indefinite in this connection confuses the theory of opposition altogether.

§ 474. Of the several forms of opposition contradictory opposition is logically the strongest. For this three reasons may be given—

(1) Contradictory opposites differ both in quantity and in quality, whereas others differ only in one or the other.

(2) Contradictory opposites are incompatible both as to truth and falsity, whereas in other cases it is only the truth *or* falsity of the two that is incompatible.

(3) Contradictory opposition is the safest form to adopt in argument. For the contradictory opposite refutes the adversary's proposition as effectually as the contrary, and is not so liable to a counter-refutation.

§ 475. At first sight indeed contrary opposition appears stronger, because it gives a more sweeping denial to the adversary's assertion. If, for instance, some person with whom we were arguing were to lay down that 'All poets are bad logicians,' we might be tempted in the heat of controversy to maintain against him the contrary proposition 'No poets are bad logicians.' This would certainly be a more emphatic contradiction, but, logically considered, it would not be as sound a one as the less obtrusive contradictory, 'Some poets are not bad logicians,' which it would be very difficult to refute.

§ 476. The phrase 'diametrically opposed to one another' seems to be one of the many expressions which have crept into common language from the technical usage of logic. The propositions A and O and E and I respectively are diametrically opposed to one another in the sense that the straight lines connecting them constitute the diagonals of the parallelogram in the scheme of opposition.

§ 477. It must be noticed that in the case of a singular proposition there is only one mode of contradiction possible. Since the quantity of such a proposition is at the minimum, the contrary and contradictory are necessarily merged into one. There is no way of denying the proposition 'This house is haunted,' save by maintaining the proposition which differs from it only in quality, namely, 'This house is not haunted.'

478. A kind of generality might indeed he imparted even to a singular proposition by expressing it in the form 'A is always B.' Thus we may say, 'This man is always idle'—a proposition which admits of being contradicted under the form 'This man is sometimes not idle.'

CHAPTER IV.

Of Conversion.

§ 479. Conversion is an immediate inference grounded On the transposition of the subject and predicate of a proposition.

§ 480. In this form of inference the antecedent is technically known as the Convertend, i.e. the proposition to be converted, and the consequent as the Converse, i.e. the proposition which has been converted.

§ 481. In a loose sense of the term we may be said to have converted a proposition when we have merely transposed the subject and predicate, when, for instance, we turn the proposition 'All A is B' into 'All B is A' or 'Some A is not B' into 'Some B is not A.' But these propositions plainly do not follow from the former ones, and it is only with conversion as a form of inference—with Illative Conversion as it is called—that Logic is concerned.

§ 482. For conversion as a form of inference two rules have been laid down—

(1) No term must be distributed in the converse which was not distributed in the convertend.

(2) The quality of the converse must be the same as that of the convertend.

§ 483. The first of these rules is founded on the nature of things. A violation of it involves the fallacy of arguing from part of a term to the whole.

§ 484. The second rule is merely a conventional one. We may make a valid inference in defiance of it: but such an inference will be seen presently to involve something more than mere conversion.

§ 485. There are two kinds of conversion—

(1) Simple.

(2) Per Accidens or by Limitation.

§ 486. We are said to have simply converted a proposition when the quantity remains the same as before.

§ 487. We are said to have converted a proposition per accidens, or by limitation, when the rules for the distribution of terms necessitate a reduction in the original quantity of the proposition.

§ 488.

A can only be converted per accidens.

E and I can be converted simply.

O cannot be converted at all.

§ 489. The reason why A can only be converted per accidens is that, being affirmative, its predicate is undistributed (§ 293). Since 'All A is B' does not mean more than 'All A is some B,' its proper converse is 'Some B is A.' For, if we endeavoured to elicit the inference, 'All B is A,' we should be distributing the term B in the converse, which was not distributed in the convertend. Hence we should be involved in the fallacy of arguing from the part to the whole. Because 'All doctors are men' it by no means follows that 'All men are doctors.'

§ 499. E and I admit of simple conversion, because the quantity of the subject and predicate is alike in each, both subject and predicate being distributed in E and undistributed in I.

/ No A is B.

E <

\ .'. No B is A.

/ Some A is B.

I <

\ ∴ Some B is A.

§ 491. The reason why O cannot be converted at all is that its subject is undistributed and that the proposition is negative. Now, when the proposition is converted, what was the subject becomes the predicate, and, as the proposition must still be negative, the former subject would now be distributed, since every negative proposition distributes its predicate. Hence we should necessarily have a term distributed in the converse which was not distributed in the convertend. From 'Some men are not doctors,' it plainly does not follow that 'Some doctors are not men'; and, generally from 'Some A is not B' it cannot be inferred that 'Some B is not A,' since the proposition 'Some A is not B' admits of the interpretation that B is wholly contained in A.

[Illustration]

§ 492. It may often happen as a matter of fact that in some given matter a proposition of the form 'All B is A' is true simultaneously with 'All A is B.' Thus it is as true to say that 'All equiangular triangles are equilateral' as that 'All equilateral triangles are equiangular.' Nevertheless we are not logically warranted in inferring the one from the other. Each has to be established on its separate evidence.

§ 493. On the theory of the quantified predicate the difference between simple conversion and conversion by limitation disappears. For the quantity of a proposition is then no longer determined solely by reference to the quantity of its subject. 'All A is some B' is of no greater quantity than 'Some B is all A,' if both subject and predicate have an equal claim to be considered.

§ 494. Some propositions occur in ordinary language in which the quantity of the predicate is determined. This is especially the case when the subject is a singular term. Such propositions admit of conversion by a mere transposition of their subject and predicate, even though they fall under the form of the A proposition, e.g.

Virtue is the condition of happiness.

∴ The condition of happiness is virtue.

And again,

Virtue is a condition of happiness.

∴ A condition of happiness is virtue.

In the one case the quantity of the predicate is determined by the form of the expression as distributed, in the other as undistributed.

§ 495. Conversion offers a good illustration of the principle on which we have before insisted, namely, that in the ordinary form of proposition the subject is used in extension and the predicate in intension. For when by conversion we change the predicate into the subject, we are often obliged to attach a noun substantive to the predicate, in order that it may be taken in extension, instead of, as before, in intension, e.g.

Some mothers are unkind.

∴ Some unkind persons are mothers.

Again,

Virtue is conducive to happiness.

∴ One of the things which are conducive to happiness is virtue.

CHAPTER V.

Of Permutation.

§ 496. Permutation [Footnote: Called by some writers Obversion.] is an immediate inference grounded on a change of quality in a proposition and a change of the predicate into its contradictory-term.

§ 497. In less technical language we may say that permutation is expressing negatively what was expressed affirmatively and vice versâ.

§ 498. Permutation is equally applicable to all the four forms of proposition.

(A) All A is B.

∴. No A is not-B (E).

(E) No A is B.

∴. All A is not-B (A).

(I) Some A is B.

∴. Some A is not not-B (O).

(O) Some A is not B.

∴. Some A is not-B (I).

§ 499, Or, to take concrete examples—

(A) All men are fallible.

∴. No men are not-fallible (E).

(E) No men are perfect.

∴. All men are not-perfect (A).

(I) Some poets are logical.

∴. Some poets are not not-logical (O).

(O) Some islands are not inhabited.

∴. Some islands are not-inhabited (I).

§ 500. The validity of permutation rests on the principle of excluded middle, namely— That one or other of a pair of contradictory terms must be applicable to a given subject, so that, when one may be predicated affirmatively, the other may be predicated negatively, and vice versâ (§ 31).

§ 501. Merely to alter the quality of a proposition would of course affect its meaning; but when the predicate is at the same time changed into its contradictory term, the original meaning of the proposition is retained, whilst the form alone is altered. Hence we may lay down the following practical rule for permutation—

Change the quality of the proposition and change the predicate into its contradictory term.

§ 502. The law of excluded middle holds only with regard to contradictories. It is not true of a pair of positive and privative terms, that one or other of them must be applicable to any given subject. For the subject may happen to fall wholly outside the sphere to which such a pair of terms is limited. But since the fact of a term being applied is a sufficient indication of its applicability, and since within a given sphere positive and privative terms are as mutually destructive as contradictories, we may in all cases substitute the privative for the negative term in immediate inference by permutation, which will bring the inferred proposition more into conformity with the ordinary usage of language. Thus the concrete instances given above will appear as follows—

(A) All men are fallible.

∴. No men are infallible (E).

(E) No men are perfect.

∴. All men are imperfect (A).

(I) Some poets are logical.

∴ Some poets are not illogical (O).

(O) Some islands are not inhabited.

∴ Some islands are uninhabited (I).

CHAPTER VI.

Of Compound Forms of Immediate Inference.

§ 503. Having now treated of the three simple forms of immediate inference, we go on to speak of the compound forms, and first of

Conversion by Negation.

§ 504. When A and O have been permuted, they become respectively E and I, and, in this form, admit of simple conversion. We have here two steps of inference: but the process may be performed at a single stroke, and is then known as Conversion by Negation. Thus from 'All A is B' we may infer 'No not-B is A,' and again from 'Some A is not B' we may infer 'Some not-B is A.' The nature of these inferences will be seen better in concrete examples.

§ 505.

(A) All poets are imaginative.

∴ No unimaginative persons are poets (E).

(O) Some parsons are not clerical.

∴ Some unclerical persons are parsons (I).

§ 506. The above inferences, when analysed, will be found to resolve themselves into two steps, namely,

(1) Permutation.

(2) Simple Conversion.

(A) All A is B.

∴ No A is not-B (by permutation).

∴ No not-B is A (by simple conversion).

(O) Some A is not B.

∴ Some A is not-B (by permutation).

∴ Some not-B is A (by simple conversion).

§ 507. The term conversion by negation has been arbitrarily limited to the exact inferential procedure of permutation followed by simple conversion. Hence it necessarily applies only to A and 0 propositions, since these when permuted become E and 1, which admit of simple conversion; whereas E and 1 themselves are permuted into A and 0, which do not. There seems to be no good reason, however, why the term 'conversion by negation' should be thus restricted in its meaning; instead of being extended to the combination of permutation with conversion, no matter in what order the two processes may be performed. If this is not done, inferences quite as legitimate as those which pass under the title of conversion by negation are left without a name.

§ 508. From E and 1 inferences may be elicited as follows—

(E) No A is B.

∴ All B is not-A (A).

(I) Some A is B.

∴ Some B is not not-A (O).

(E) No good actions are unbecoming.

∴. All unbecoming actions are not-good (A).

(I) Some poetical persons are logicians.

∴. Some logicians are not unpoetical (O).

Or, taking a privative term for our subject,

Some unpractical persons are statesmen.

∴. Some statesmen are not practical.

§ 509. When the inferences just given are analysed, it will be found that the process of simple conversion precedes that of permutation.

§ 510. In the case of the E proposition a compound inference can be drawn even in the original order of the processes,

No A is B.

∴. Some not-B is A.

No one who employs bribery is honest.

∴. Some dishonest men employ bribery.

The inference here, it must be remembered, does not refer to matter of fact, but means that one of the possible forms of dishonesty among men is that of employing bribery.

§ 511. If we analyse the preceding, we find that the second step is conversion by limitation.

No A is B.

∴. All A is not-B (by permutation).

∴. Some not-B is A (by conversion per accidens).

§ 512. From A again an inference can be drawn in the reverse order of conversion per accidens followed by permutation—

All A is B.

∴. Some B is not not-A.

All ingenuous persons are agreeable.

∴. Some agreeable persons are not disingenuous.

§ 513. The intermediate link between the above two propositions is the converse per accidens of the first—'Some B is A.' This inference, however, coincides with that from 1 (§ 508), as the similar inference from E (§ 510) coincides with that from 0 (§ 506).

§ 514. All these inferences agree in the essential feature of combining permutation with conversion, and should therefore be classed under a common name.

§ 515. Adopting then this slight extension of the term, we define conversion by negation as—A form of conversion in which the converse differs in quality from the convertend, and has the contradictory of one of the original terms.

§ 516. A still more complex form of immediate inference is known as *Conversion by Contraposition.*

This mode of inference assumes the following form—

All A is B.

∴. All not-B is not-A.

All human beings are fallible.

∴. All infallible beings are not-human.

§ 517. This will be found to resolve itself on analysis into three steps of inference in the following order—

(1) Permutation.

(2) Simple Conversion.

(3) Permutation.

§ 518. Let us verify this statement by performing the three steps.

All A is B.

∴. No A is not-B (by permutation).

∴ No not-B is A (by simple conversion).

∴ All not-B is not-A (by permutation).

All Englishmen are Aryans.

∴ No Englishmen are non-Aryans.

∴ No non-Aryans are Englishmen.

∴ All non-Aryans are non-Englishmen.

§ 519. Conversion by contraposition may be complicated in appearance by the occurrence of a negative term in the subject or predicate or both, e.g.

All not-A is B.

∴ All not-B is A.

Again,

All A is not-B.

∴ All B is not-A.

Lastly,

All not-A is not-B.

∴ All B is A.

§ 520. The following practical rule will be found of use for the right performing of the process—

Transpose the subject and predicate, and substitute for each its contradictory term.

§ 521. As concrete illustrations of the above forms of inference we may take the following—

All the men on this board that are not white are red.

∴ All the men On this board that are not red are white.

Again,

All compulsory labour is inefficient.

∴ All efficient labour is free (=non-compulsory).

Lastly,

All inexpedient acts are unjust.

∴ All just acts are expedient.

§ 522. Conversion by contraposition may be said to rest on the following principle—

If one class be wholly contained in another, whatever is external to the containing class is external also to the class contained.

[Illustration]

§ 523. The same principle may be expressed intensively as follows:—

If an attribute belongs to the whole of a subject, whatever fails to exhibit that attribute does not come under the subject.

§ 524. This statement contemplates conversion by contraposition only in reference to the A proposition, to which the process has hitherto been confined. Logicians seem to have overlooked the fact that conversion by contraposition is as applicable to the O as to the A proposition, though, when expressed in symbols, it presents a more clumsy appearance.

Some A is not B.

∴ Some not-B is not not-A.

Some wholesome things are not pleasant.

∴ Some unpleasant things are not unwholesome.

§ 525. The above admits of analysis in exactly the same way as the same process when applied to the A proposition.

Some A is not B.

∴ Some A is not-B (by permutation).

∴ Some not-B is A (by simple conversion).

∴ Some not-B is not not-A (by permutation).

The result, as in the case of the A proposition, is the converse by negation of the original proposition permuted.

§ 526. Contraposition may also be applied to the E proposition by the use of conversion per accidens in the place of simple conversion. But, owing to the limitation of quantity thus effected, the result arrived at is the same as in the case of the O proposition. Thus from 'No wholesome things are pleasant' we could draw the same inference as before. Here is the process in symbols, when expanded.

No A is B.

∴. All A is not-B (by permutation).

∴. Some not-B is A (by conversion per accidens).

∴. Some not-B is not not-A (by permutation).

§ 527. In its unanalysed form conversion by contraposition may be defined generally as—A form of conversion in which both subject and predicate are replaced by their contradictories.

§ 528. Conversion by contraposition differs in several respects from conversion by negation.

(1) In conversion by negation the converse differs in quality from the convertend: whereas in conversion by contraposition the quality of the two is the same.

(2) In conversion by negation we employ the contradictory either of the subject or predicate, but in conversion by contraposition we employ the contradictory of both.

(3) Conversion by negation involves only two steps of immediate inference: conversion by contraposition three.

§ 529. Conversion by contraposition cannot be applied to the ordinary E proposition except by limitation (§ 526).

From 'No A is B' we cannot infer 'No not-B is not-A.' For, if we could, the contradictory of the latter, namely, 'Some not-B is not-A' would be false. But it is manifest that this is not necessarily false. For when one term is excluded from another, there must be numerous individuals which fall under neither of them, unless it should so happen that one of the terms is the direct contradictory of the other, which is clearly not conveyed by the form of the expression 'No A is B. 'No A is not-A' stands alone among E propositions in admitting of full conversion by contraposition, and the form of that is the same after it as before.

§ 530. Nor can conversion by contraposition be applied at all to I.

[Illustration]

From 'Some A is B' we cannot infer that 'Some not-B is not-A.' For though the proposition holds true as a matter of fact, when A and B are in part mutually exclusive, yet this is not conveyed by the form of the expression. It may so happen that B is wholly contained under A, while A itself contains everything. In this case it will be true that 'No not-B is not-A,' which contradicts the attempted inference. Thus from the proposition 'Some things are substances' it cannot be inferred that 'Some not-substances are not-things,' for in this case the contradictory is true that 'No not-substances are not-things'; and unless an inference is valid in every case, it is not formally valid at all.

§ 531. It should be noticed that in the case of the [nu] proposition immediate inferences are possible by mere contraposition without conversion.

All A is all B.

∴. All not-A is not-B.

For example, if all the equilateral triangles are all the equiangular, we know at once that all non-equilateral triangles are also non-equiangular.

§ 532. The principle upon which this last kind of inference rests is that when two terms are co-extensive, whatever is excluded from the one is excluded also from the other.

CHAPTER VII.

Of other Forms of Immediate Inference.

§ 533. Having treated of the main forms of immediate inference, whether simple or compound, we will now close this subject with a brief allusion to some other forms which have been recognised by logicians.

§ 534. Every statement of a relation may furnish us with ail immediate inference in which the same fact is presented from the opposite side. Thus from 'John hit James' we infer 'James was hit by John'; from 'Dick is the grandson of Tom' we infer 'Tom is the grandfather of Dick'; from 'Bicester is north-east of Oxford' we infer 'Oxford is south-west of Bicester'; from 'So and so visited the Academy the day after he arrived in London' we infer 'So and so arrived in London the day before he visited the Academy'; from 'A is greater than B' we infer 'B is less than A'; and so on without limit. Such inferences as these are material, not formal. No law can be laid down for them except the universal postulate, that

'Whatever is true in one form of words is true in every other form of words which conveys the same meaning.'

§ 535. There is a sort of inference which goes under the title of Immediate Inference by Added Determinants, in which from some proposition already made another is inferred, in which the same attribute is attached both to the subject and the predicate, e.g.,

A horse is a quadruped.

∴ A white horse is a white quadruped.

§ 536. Such inferences are very deceptive. The attributes added must be definite qualities, like whiteness, and must in no way involve a comparison. From 'A horse is a quadruped' it may seem at first sight to follow that 'A swift horse is a swift quadruped.' But we need not go far to discover how little formal validity there is about such an inference. From 'A horse is a quadruped' it by no means follows that 'A slow horse is a slow quadruped'; for even a slow horse is swift compared with most quadrupeds. All that really follows here is that 'A slow horse is a quadruped which is slow for a horse.' Similarly, from 'A Bushman is a man' it does not follow that 'A tall Bushman is a tall man,' but only that 'A tall Bushman is a man who is tall for a Bushman'; and so on generally.

§ 537. Very similar to the preceding is the process known as Immediate Inference by Complex Conception, e.g.

A horse is a quadruped.

∴ The head of a horse is the head of a quadruped.

§ 538. This inference, like that by added determinants, from which it differs in name rather than in nature, may be explained on the principle of Substitution. Starting from the identical proposition, 'The head of a quadruped is the head of a quadruped,' and being given that 'A horse is a quadruped,' so that whatever is true of 'quadruped' generally we know to be true of 'horse,' we are entitled to substitute the narrower for the wider term, and in this manner we arrive at the proposition,

The head of a horse is the head of a quadruped.

§ 539. Such an inference is valid enough, if the same caution be observed as in the case of added determinants, that is, if no difference be allowed to intervene in the relation of the fresh conception to the generic and the specific terms.

CHAPTER VIII.

Of Mediate Inferences or Syllogisms.

§ 540. A Mediate Inference, or Syllogism, consists of two propositions, which are called the Premisses, and a third proposition known as the Conclusion, which flows from the two conjointly.

§ 541. In every syllogism two terms are compared with one another by means of a third, which is called the Middle Term. In the premisses each of the two terms is compared separately with the middle term; and in the conclusion they are compared with one another.

§ 542. Hence every syllogism consists of three terms, one of which occurs twice in the premisses and does not appear at all in the conclusion. This term is called the Middle Term. The predicate of the conclusion is called the Major Term and its subject the Minor Term.

§ 543. The major and minor terms are called the Extremes, as opposed to the Mean or Middle Term.

§ 544. The premiss in which the major term is compared with the middle is called the Major Premiss.

§ 545. The other premiss, in which the minor term is compared with the middle, is called the Minor Premiss.

§ 546. The order in which the premisses occur in a syllogism is indifferent, but it is usual, for convenience, to place the major premiss first.

§ 547. The following will serve as a typical instance of a syllogism—

Middle term Major term \
Major Premiss. All mammals are warm-blooded | Antecedent
> or
Minor term Middle term | Premisses
Minor Premiss. All whales are mammals /
Minor term Major term \ Consequent or
∴ All whales are warm-blooded > Conclusion.

§ 548. The reason why the names 'major, 'middle' and 'minor' terms were originally employed is that in an affirmative syllogism such as the above, which was regarded as the perfect type of syllogism, these names express the relative quantity in extension of the three terms.

[Illustration]

§ 549. It must be noticed however that, though the middle term cannot be of larger extent than the major nor of smaller extent than the minor, if the latter be distributed, there is nothing to prevent all three, or any two of them, from being coextensive.

§ 550. Further, when the minor term is undistributed, we either have a case of the intersection of two classes, from which it cannot be told which of them is the larger, or the minor term is actually larger than the middle, when it stands to it in the relation of genus to species, as in the following syllogism—

All Negroes have woolly hair.
Some Africans are Negroes.
∴ Some Africans have woolly hair.

[Illustration]

§ 551. Hence the names are not applied with strict accuracy even in the case of the affirmative syllogism; and when the syllogism is negative, they are not applicable at all: since in negative propositions we have no means of comparing the relative extension of the terms employed. Had we said in the major premiss of our typical syllogism, 'No mammals are cold-blooded,' and drawn the conclusion 'No whales are cold-blooded,' we could not have compared the relative extent of the terms 'mammal' and 'cold-blooded,' since one has been simply excluded from the other.

[Illustration]

§ 552. So far we have rather described than defined the syllogism. All the products of thought, it will be remembered, are the results of comparison. The syllogism, which is one of them, may be so regarded in two ways—

(1) As the comparison of two propositions by means of a third.

(2) As the comparison of two terms by means of a third or middle term.

§ 553. The two propositions which are compared with one another are the major premiss and the conclusion, which are brought into connection by means of the minor premiss. Thus in the syllogism above given we compare the conclusion 'All whales are warm-blooded' with the major premiss 'All mammals are warm-blooded,' and find that the former is contained under the latter, as soon as we become acquainted with the intermediate proposition 'All whales are mammals.'

§ 554. The two terms which are compared with one another are of course the major and minor.

§ 555. The syllogism is merely a form into which our deductive inferences may be thrown for the sake of exhibiting their conclusiveness. It is not the form which they naturally assume in speech or writing. Practically the conclusion is generally stated first and the premisses introduced by some causative particle as 'because,' 'since,' 'for,' &c. We start with our conclusion, and then give the reason for it by supplying the premisses.

§ 556. The conclusion, as thus stated first, was called by logicians the Problema or Quaestio, being regarded as a problem or question, to which a solution or answer was to be found by supplying the premisses.

§ 557. In common discourse and writing the syllogism is usually stated defectively, one of the premisses or, in some cases, the conclusion itself being omitted. Thus instead of arguing at full length

All men are fallible,

The Pope is a man,

.'. The Pope is fallible,

we content ourselves with saying 'The Pope is fallible, for he is a man,' or 'The Pope is fallible, because all men are so'; or perhaps we should merely say 'All men are fallible, and the Pope is a man,' leaving it to the sagacity of our hearers to supply the desired conclusion. A syllogism, as thus elliptically stated, is commonly, though incorrectly, called an Enthymeme. When the major premiss is omitted, it is called an Enthymeme of the First Order; when the minor is omitted, an Enthymeme of the Second Order; and when the conclusion is omitted an Enthymeme of the Third Order.

CHAPTER IX.

Of Mood and Figure.

§ 558. Syllogisms may differ in two ways—

(1) in Mood;

(2) in Figure.

§ 559. Mood depends upon the kind of propositions employed. Thus a syllogism consisting of three universal affirmatives, AAA, would be said to differ in mood from one consisting of such propositions as EIO or any other combination that might be made. The syllogism previously given to prove the fallibility of the Pope belongs to the mood AAA. Had we drawn only a particular conclusion, 'Some Popes are fallible,' it would have fallen into the mood AAI.

§ 560. Figure depends upon the arrangement of the terms in the propositions. Thus a difference of figure is internal to a difference of mood, that is to say, the same mood can be in any figure.

§ 561. We will now show how many possible varieties there are of mood and figure, irrespective of their logical validity.

§ 562. And first as to mood.

Since every syllogism consists of three propositions, and each of these propositions may be either A, E, I, or O, it is clear that there will be as many possible moods as there can be combinations of four things, taken three together, with no restrictions as to repetition. It will be seen that there are just sixty-four of such combinations. For A may be followed either by itself or by E, I, or O. Let us suppose it to be followed by itself. Then this pair of premisses, AA, may have for its conclusion either A, E, I, or O, thus giving four combinations which commence with AA. In like manner there will be four commencing with AE, four with AI, and four with AO, giving a total of sixteen combinations which commence with A. Similarly there will be sixteen commencing with E, sixteen with I, sixteen with O—in all sixty-four. It is very few, however, of these possible combinations that will be found legitimate, when tested by the rules of syllogism.

§ 563. Next as to figure.

There are four possible varieties of figure in a syllogism, as may be seen by considering the positions that can be occupied by the middle term in the premisses. For as there are only two terms in each premiss, the position occupied by the middle term necessarily determines that of the others. It is clear that the middle term must either occupy the same position in both premisses or not, that is, it must either be subject in both or predicate in both, or else subject in one and predicate in the other. Now, if we are not acquainted with the conclusion of our syllogism, we do not know which is the major and which the minor term, and have therefore no means of distinguishing between one premiss and another; consequently we must Stop here, and say that there are only three different arrangements possible. But, if the Conclusion also be assumed as known, then we are able to distinguish one premiss as the major and the other as the minor; and so we can go further, and lay down that, if the middle term does not hold the same position in both premisses, it must either be subject in the major and predicate in the minor, or else predicate in the major and subject in the minor.

§ 564. Hence there result

The Four Figures.

When the middle term is subject in the major and predicate in the minor, we are said to have the First Figure.

When the middle term is predicate in both premisses, we are said to have the Second Figure.

When the middle term is subject in both premisses, we are said to have the Third Figure.

When the middle term is predicate in the major premiss and subject in the minor, we are said to have the Fourth Figure.

§ 565. Let A be the major term; B the middle. C the minor.

Figure I. Figure II. Figure III. Figure IV.

B—A A—B B—A A—B
C—B C—B B—C B—C
C—A C—A C—A C—A

All these figures are legitimate, though the fourth is comparatively valueless.

§ 566. It will be well to explain by an instance the meaning of the assertion previously made, that a difference of figure is internal to a difference of mood. We will take the mood EIO, and by varying the position of the terms, construct a syllogism in it in each of the four figures.

I.

E No wicked man is happy.

I Some prosperous men are wicked.

O .'. Some prosperous men are not happy.

II.

E No happy man is wicked.

I Some prosperous men are wicked.

O .'. Some prosperous men are not happy.

III.

E No wicked man is happy.

I Some wicked men are prosperous.

O .'. Some prosperous men are not happy.

IV.

E No happy man is wicked.

I Some wicked men are prosperous.

O .'. Some prosperous men are not happy.

§ 567. In the mood we have selected, owing to the peculiar nature of the premisses, both of which admit of simple conversion, it happens that the resulting syllogisms are all valid. But in the great majority of moods no syllogism would be valid at all, and in many moods a syllogism would be valid in one figure and invalid in another. As yet however we are only concerned with the conceivable combinations, apart from the question of their legitimacy.

§ 568. Now since there are four different figures and sixty-four different moods, we obtain in all 256 possible ways of arranging three terms in three propositions, that is, 256 possible forms of syllogism.

CHAPTER X.

Of the Canon of Reasoning.

& 569. The first figure was regarded by logicians as the only perfect type of syllogism, because the validity of moods in this figure may be tested directly by their complying, or failing to comply, with a certain axiom, the truth of which is self-evident. This axiom is known as the Dictum de Omni et Nullo. It may be expressed as follows—

Whatever may be affirmed or denied of a whole class may be affirmed or denied of everything contained in that class.

§ 570. This mode of stating the axiom contemplates predication as being made in extension, whereas it is more naturally to be regarded as being made in intension.

§ 571. The same principle may be expressed intensively as follows—

Whatever has certain attributes has also the attributes which invariably accompany them .[Footnote: Nota notae est nota rei ipsius. 'Whatever has any mark has that which it is a mark of.' Mill, vol. i, p. 201,]

§ 572. By Aristotle himself the principle was expressed in a neutral form thus—

'Whatever is stated of the predicate will be stated also of the subject [Footnote: [Greek: osa katà toû kategorouménou légetai pánta kaì katà toû hypokeiménou rhaetésetai]. Cat. 3, § I].'

This way of putting it, however, is too loose.

§ 573. The principle precisely stated is as follows—

Whatever may be affirmed or denied universally of the predicate of an affirmative proposition, may be affirmed or denied also of the subject.

§ 574. Thus, given an affirmative proposition 'Whales are mammals,' if we can affirm anything universally of the predicate 'mammals,' as, for instance, that 'All mammals are warm-blooded,' we shall be able to affirm the same of the subject 'whales'; and, if we can deny anything universally of the predicate, as that 'No mammals are oviparous,' we shall be able to deny the same of the subject.

§ 575. In whatever way the supposed canon of reasoning may be stated, it has the defect of applying only to a single figure, namely, the first. The characteristic of the reasoning in that figure is that some general rule is maintained to hold good in a particular case. The major premiss lays down some general principle, whether affirmative or negative; the minor premiss asserts that a particular case falls under this principle; and the conclusion applies the general principle to the particular case. But though all syllogistic reasoning may be tortured into conformity with this type, some of it finds expression more naturally in other ways.

§ 576. Modern logicians therefore prefer to abandon the Dictum de Omni et Nullo in any shape, and to substitute for it the following three axioms, which apply to all figures alike.

Three Axioms of Mediale Inference.

(1) If two terms agree with the same third term, they agree with one another.

(2) If one term agrees, and another disagrees, with the same third term, they disagree with one another.

(3) If two terms disagree with the same third term, they may or may not agree with one another.

§ 577. The first of these axioms is the principle of all affirmative, the second of all negative, syllogisms; the third points out the conditions under which no conclusion can be drawn. If there is any agreement at all between the two terms and the third, as in the cases contemplated in the first and second axioms, then we have a conclusion of some kind: if it is otherwise, we have none.

§ 578. It must be understood with regard to these axioms that, when we speak of terms agreeing or disagreeing with the same third term, we mean that they agree or disagree with the same part of it.

§ 579. Hence in applying these axioms it is necessary to bear in mind the rules for the distinction of terms. Thus from

All B is A,

No C is B,

the only inference which can be drawn is that Some A is not C (which alters the figure from the first to the fourth). For it was only part of A which was known to agree with B. On the theory of the quantified predicate we could draw the inference No C is some A.

§ 580. It is of course possible for terms to agree with different parts of the same third term, and yet to have no connection with one another. Thus

All birds fly.

All bats fly.

But we do not infer therefrom that bats are birds or vice versâ.

§ 581. On the other hand, had we said,—

All birds lay eggs,

No bats lay eggs,

we might confidently have drawn the conclusion

No bats are birds

For the term 'bats,' being excluded from the whole of the term 'lay eggs,' is thereby necessarily excluded from that part of it which coincides with 'birds.'

[Illustration]

CHAPTER XI.

Of the Generad Rules of Syllogism.

§ 582. We now proceed to lay down certain general rules to which all valid syllogisms must conform. These are divided into primary and derivative.

I. *Primary.*

(1) A syllogism must consist of three propositions only.

(2) A syllogism must consist of three terms only.

(3) The middle term must be distributed at least once in the premises.

(4) No term must be distributed in the conclusion which was not distributed in the premises.

(5) Two negative premises prove nothing.

(6) If one premiss be negative, the conclusion must be negative.

(7) If the conclusion be negative, one of the premisses must be negative: but if the conclusion be affirmative, both premisses must be affirmative.

II. *Derivative.*

(8) Two particular premisses prove nothing.

(9) If one premiss be particular, the conclusion must be particular.

§ 583. The first two of these rules are involved in the definition of the syllogism with which we started. We said it might be regarded either as the comparison of two propositions by means of a third or as the comparison of two terms by means of a third. To violate either of these rules therefore would be inconsistent with the fundamental conception of the syllogism. The first of our two definitions indeed (§ 552) applies directly only to the syllogisms in the first figure; but since all syllogisms may be expressed, as we shall presently see, in the first figure, it applies indirectly to all. When any process of mediate inference appears to have more than two premises, it will always be found that there is more than one syllogism. If there are less than three propositions, as in the fallacy of 'begging the question,' in which the conclusion simply reiterates one of the premises, there is no syllogism at all.

With regard to the second rule, it is plain that any attempted syllogism which has more than three terms cannot conform to the conditions of any of the axioms of mediate inference.

§ 584. The next two rules guard against the two fallacies which are fatal to most syllogisms whose constitution is unsound.

§ 585. The violation of Rule 3 is known as the Fallacy of Undistributed Middle. The reason for this rule is not far to seek. For if the middle term is not used in either premiss in its whole extent, we may be referring to one part of it in one premiss and to quite another part of it in another, so that there will be really no middle term at all. From such premises as these—

All pigs are omnivorous,

All men are omnivorous,

it is plain that nothing follows. Or again, take these premisses—

Some men are fallible,

All Popes are men.

Here it is possible that 'All Popes' may agree with precisely that part of the term 'man,' of which it is not known whether it agrees with 'fallible' or not.

§ 586. The violation of Rule 4 is known as the Fallacy of Illicit Process. If the major term is distributed in the conclusion, not having been distributed in the premiss, we have what is called Illicit Process of the Major; if the same is the case with the minor term, we have Illicit Process of the Minor.

§ 587. The reason for this rule is that if a term be used in its whole extent in the conclusion, which was not so used in the premiss in which it occurred, we would be arguing from the part to the whole. It is the same sort of fallacy which we found to underlie the simple conversion of an A proposition.

§ 588. Take for instance the following—

All learned men go mad.

John is not a learned man.

∴ John will not go mad.

In the conclusion 'John' is excluded from the whole class of persons who go mad, whereas in the premisses, granting that all learned men go mad, it has not been said that they are all the men who do so. We have here an illicit process of the major term.

§ 589. Or again take the following—

All Radicals are covetous.

All Radicals are poor.

∴ All poor men are covetous.

The conclusion here is certainly not warranted by our premisses. For in them we spoke only of some poor men, since the predicate of an affirmative proposition is undistributed.

§ 590. Rule 5 is simply another way of stating the third axiom of mediate inference. To know that two terms disagree with the same third term gives us no ground for any inference as to whether they agree or disagree with one another, e.g.

Ruminants are not oviparous.

Sheep are not oviparous.

For ought that can be inferred from the premisses, sheep may or may not be ruminants.

§ 591. This rule may sometimes be violated in appearance, though not in reality. For instance, the following is perfectly legitimate reasoning.

No remedy for corruption is effectual that does not render it useless.

Nothing but the ballot renders corruption useless.

∴ Nothing but the ballot is an effectual remedy for corruption.

But on looking into this we find that there are four terms—

No not-A is B.

No not-C is A.

∴ No not-C is B.

The violation of Rule 5 is here rendered possible by the additional violation of Rule 2. In order to have the middle term the same in both premisses we are obliged to make the minor affirmative, thus

No not-A is B.

All not-C is not-A.

∴ No not-C is B.

No remedy that fails to render corruption useless is effectual.

All but the ballot fails to render corruption useless.

∴ Nothing but the ballot is effectual.

§ 592. Rule 6 declares that, if one premiss be negative, the conclusion must be negative. Now in compliance with Rule 5, if one premiss be negative, the other must be affirmative. We have therefore the case contemplated in the second axiom, namely, of one term agreeing and the other disagreeing with the same third term; and we know that this can only give ground for a judgement of disagreement between the two terms themselves—in other words, to a negative conclusion.

§ 593. Rule 7 declares that, if the conclusion be negative, one of the premisses must be negative; but, if the conclusion be affirmative, both premisses must be affirmative. It is plain from the axioms that a judgement of disagreement can only be elicited from a judgement of agreement combined with a judgement of disagreement, and that a judgement of agreement can result only from two prior judgements of agreement.

§ 594. The seven rules already treated of are evident by their own light, being of the nature of definitions and axioms: but the two remaining rules, which deal with particular premisses, admit of being proved from their predecessors.

§ 595. Proof of Rule 8.—*That two particular premisses prove nothing.*

We know by Rule 5 that both premisses cannot be negative. Hence they must be either both affirmative, II, or one affirmative and one negative, IO or OI.

Now II premisses do not distribute any term at all, and therefore the middle term cannot be distributed, which would violate Rule 3.

Again in IO or OI premisses there is only one term distributed, namely, the predicate of the O proposition. But Rule 3 requires that this one term should be the middle term. Therefore the major term must be undistributed in the major premiss. But since one of the premisses is negative, the conclusion must be negative, by Rule 6. And every negative proposition distributes its predicate. Therefore the major term must be distributed where it occurs as predicate of the conclusion. But it was not distributed in the major premiss. Therefore in drawing any conclusion we violate Rule 4 by an illicit process of the major term.

§ 596. Proof of Rule 9.—*That, if* one *premiss be particular, the conclusion must be particular.*

Two negative premisses being excluded by Rule 5, and two particular by Rule 8, the only pairs of premisses we can have are—

AI, AO, EI.

Of course the particular premiss may precede the universal, but the order of the premisses will not affect the reasoning.

AI premisses between them distribute one term only. This must be the middle term by Rule 3. Therefore the conclusion must be particular, as its subject cannot be distributed,

AO and EI premisses each distribute two terms, one of which must be the middle term by Rule 3: so that there is only one term left which may be distributed in the conclusion. But the conclusion must be negative by Rule 4. Therefore its predicate must be distributed. Hence its subject cannot be so. Therefore the conclusion must be particular.

§ 597. Rules 6 and 9 are often lumped together in a single expression—'The conclusion must follow the weaker part,' negative being considered weaker than affirmative, and particular than universal.

§ 598. The most important rules of syllogism are summed up in the following mnemonic lines, which appear to have been perfected, though not invented, by a mediæval logician known as Petrus Hispanus, who was afterwards raised to the Papal Chair under the title of Pope John XXI, and who died in 1277—

Distribuas medium, nec quartus terminus adsit;
Utraque nec praemissa negans, nec particularis;
Sectetur partem conclusio deteriorem,
Et non distribuat, nisi cum praemissa, negetve.

CHAPTER XII.

Of the Determination of the Legitimate Moods of Syllogism.

§ 599. It will be remembered that there were found to be 64 possible moods, each of which might occur in any of the four figures, giving us altogether 256 possible varieties of syllogism. The task now before us is to determine how many of these combinations of mood and figure are legitimate.

§ 600. By the application of the preceding rules we are enabled to reduce the 64 possible moods to 11 valid ones. This may be done by a longer or a shorter method. The longer method, which is perhaps easier of comprehension, is to write down the 64 possible moods, and then strike out such as violate any of the rules of syllogism.

AAA -AEA- -AIA- -AOA- -AAE- AEE -AIE- -AOE- AAI -AEI- AII -AOI- -AAO- AEO -AIO- AOO

-EAA- -EEA- -EIA- -EOA- EAE -EEE- -EIE- -EOE- -EAI- -EEI- -EII- -EOI- EAO -EEO- EIO -EOO-

[Illustration]

§ 601. The batches which are crossed are those in which the premisses can yield no conclusion at all, owing to their violating Rule 6 or 9; in the rest the premisses are legitimate, but a wrong conclusion is drawn from each of them as are translineated.

§ 602. IEO stands alone, as violating Rule 4. This may require a little explanation.

Since the conclusion is negative, the major term, which is its predicate, must be distributed. But the major premiss, being 1, does not distribute either subject or predicate. Hence IEO must always involve an illicit process of the major.

§ 603. The II moods which have been left valid, after being tested by the syllogistic rules, are as follows—

AAA. AAI. AEE. AEO. AII. AOO. EAE. EAO. EIO. IAI. OAO.

§ 604. We will now arrive at the same result by a shorter and more scientific method. This method consists in first determining what pairs of premisses are valid in accordance with Rules 6 and g, and then examining what conclusions may be legitimately inferred from them in accordance with the other rules of syllogism.

§ 605. The major premiss may be either A, E, I or O. If it is A, the minor also may be either A, E, I or O. If it is E, the minor can only be A or I. If it is I, the minor can only be A or E. If it is O, the minor can only be A. Hence there result 9 valid pairs of premisses.

AA. AE. AI. AO. EA. EI. IA. IE. OA.

Three of these pairs, namely AA, AE, EA, yield two conclusions apiece, one universal and one particular, which do not violate any of the rules of syllogism; one of them, IE, yields no conclusion at all; the remaining five have their conclusion limited to a single proposition, on the principle that the conclusion must follow the weaker part. Hence we arrive at the same result as before, of II legitimate moods—

AAA. AAI. AEE. AEO. EAE. EAO. AII. AOO. EIO. IAI. OAO.

CHAPTER XIII.

Of the Special Rules of the Four Figures.

§ 606. Our next task must be to determine how far the 11 moods which we arrived at in the last chapter are valid in the four figures. But before this can be done, we must lay down the

Special Rules of the Four Figures.

FIGURE 1.

Rule 1, The minor premiss must be affirmative.

Rule 2. The major premiss must be universal.

FIGURE II.

Rule 1. One or other premiss must be negative.

Rule 2. The conclusion must be negative.

Rule 3. The major premiss must be universal.

FIGURE III.

Rule 1. The minor premiss must be affirmative.

Rule 2. The conclusion must be particular.

FIGURE IV.

Rule 1. When the major premiss is affirmative, the minor must be universal.

Rule 2. When the minor premiss is particular, the major must be negative.

Rule 3, When the minor premiss is affirmative, the conclusion must be particular.

Rule 4. When the conclusion is negative, the major premiss must be universal.

Rule 5. The conclusion cannot be a universal affirmative.

Rule 6. Neither of the premisses can be a particular negative.

§ 607. The special rules of the first figure are merely a reassertion in another form of the Dictum de Omni et Nullo. For if the major premiss were particular, we should not have anything affirmed or denied of a whole class; and if the minor premiss were negative, we should not have anything declared to be contained in that class. Nevertheless these rules, like the rest, admit of being proved from the position of the terms in the figure, combined with the rules for the distribution of terms (§ 293).

Proof of the Special Rules of the Four Figures.

FIGURE 1.

§ 608. Proof of Rule 1.—*The minor premiss must be affirmative.*

B—A C—B C—A

If possible, let the minor premiss be negative. Then the major must be affirmative (by Rule 5), [Footnote: This refers to the General Rules of Syllogism.] and the conclusion must be negative (by Rule 6). But the major being affirmative, its predicate is undistributed; and the conclusion being negative, its predicate is distributed. Now the major term is in this figure predicate both in the major premiss and in the conclusion. Hence there results illicit process of the major term. Therefore the minor premiss must be affirmative.

§ 609. Proof of Rule 2.—*The major premiss must be universal.*

Since the minor premiss is affirmative, the middle term, which is its predicate, is undistributed there. Therefore it must be distributed in the major premiss, where it is subject. Therefore the major premiss must be universal.

FIGURE II.

§ 610. Proof of Rule 1,—*One or other premiss must be negative.*

A—B C—B C—A

The middle term being predicate in both premisses, one or other must be negative; else there would be undistributed middle.

§ 611. Proof of Rule 2.—*The conclusion must be negative.*

Since one of the premisses is negative, it follows that the conclusion also must be so (by Rule 6).

§ 612. Proof of Rule 3.—*The major premiss must be universal.*

The conclusion being negative, the major term will there be distributed. But the major term is subject in the major premiss. Therefore the major premiss must be universal (by Rule 4).

FIGURE III.

§ 613. Proof of Rule 1.—*The minor premiss must be affirmative.*

B—A B—C C—A

The proof of this rule is the same as in the first figure, the two figures being alike so far as the major term is concerned.

§ 614. Proof of Rule 2.—*The conclusion must be particular.*

The minor premiss being affirmative, the minor term, which is its predicate, will be undistributed there. Hence it must be undistributed in the conclusion (by Rule 4). Therefore the conclusion must be particular.

FIGURE IV.

§ 615. Proof of Rule I.—*When the major premiss is affirmative, the minor must be universal.*

If the minor were particular, there would be undistributed middle. [Footnote: Shorter proofs are employed in this figure, as the student is by this time familiar with the method of procedure.]

§ 616. Proof of Rule 2.—*When the minor premiss is particular, the major must be negative.*

A—B B—C C—A

This rule is the converse of the preceding, and depends upon the same principle.

§ 617. Proof of Rule 3.—*When the minor premiss is affirmative, the conclusion must be particular.*

If the conclusion were universal, there would be illicit process of the minor.

§ 618. Proof of Rule 4.—*When the conclusion is negative, the major premiss must* be universal.

If the major premiss were particular, there would be illicit process of the major.

§ 619. Proof of Rule 5.—*The conclusion CANNOT be A UNIVERSAL affirmative.*

The conclusion being affirmative, the premisses must be so too (by Rule 7). Therefore the minor term is undistributed in the minor premiss, where it is predicate. Hence it cannot be distributed in the conclusion (by Rule 4). Therefore the affirmative conclusion must be particular.

§ 620. Proof of Rule 6.—*Neither of the premisses can lie a, PARTICULAR NEGATIVE.*

If the major premiss were a particular negative, the conclusion would be negative. Therefore the major term would be distributed in the conclusion. But the major premiss being particular, the major term could not be distributed there. Therefore we should have an illicit process of the major term.

If the minor premiss were a particular negative, then, since the major must be affirmative (by Rule 5), we should have undistributed middle.

CHAPTER XIV

Of the Determination of the Moods that are valid in the Four Figures.

§ 621. By applying the special rules just given we shall be able to determine how many of the eleven legitimate moods are valid in the four figures.

$622. These eleven legitimate moods were found to be

AAA. AAI. AEE. AEO. AII. AOO. EAE. EAO. EIO. IAI. OAO.

FIGURE 1.

§ 623. The rule that the major premiss must be universal excludes the last two moods, IAI, OAO. The rule that the minor premiss must be affirmative excludes three more, namely, AEE, AEO, AOO.

Thus we are left with six moods which are valid in the first figure, namely,

AAA. EAE. AII. EIO. AAI. EAO.

FIGURE II.

§ 624. The rule that one premiss must be negative excludes four moods, namely, AAA, AAI, AII, IAI. The rule that the major must be universal excludes OAO. Thus we are left with six moods which are valid in the second figure, namely,

EAE. AEE. EIO. AOO. EAO. AEO.

FIGURE III.

§ 625. The rule that the conclusion must be particular confines us to eight moods, two of which, namely AEE and AOO, are excluded by the rule that the minor premiss must be affirmative.

Thus we are left with six moods which are valid in the third figure, namely,

AAI. IAI. AII. EAO. OAO. EIO.

FIGURE IV.

§ 626. The first of the eleven moods, AAA, is excluded by the rule that the conclusion cannot be a universal affirmative.

Two more moods, namely AOO and OAO, are excluded by the rule that neither of the premisses can be a particular negative.

AII violates the rule that when the major premiss is affirmative, the minor must be universal.

EAE violates the rule that, when the minor premiss is affirmative, the conclusion must be particular. Thus we are left with six moods which are valid in the fourth figure, namely,

AAI. AEE. IAI. EAO. EIO. AEO.

§ 627. Thus the 256 possible forms of syllogism have been reduced to two dozen legitimate combinations of mood and figure, six moods being valid in each of the four figures.

FIGURE I. AAA. EAE. AII. EIO. (AAI. EAO.)
FIGURE II. EAE. AEE. EIO. AGO. (EAO. AEO.)
FIGURE III. AAI. IAI. AII. EAO. OAO. EIO.
FIGURE IV. AAI. AEE. IAI. EAO. EIO. (AEO.)

§ 628. The five moods enclosed in brackets, though valid, are useless. For the conclusion drawn is less than is warranted by the premisses. These are called Subaltern Moods, because their conclusions might be inferred by subalternation from the universal conclusions which can justly be drawn from the same premisses. Thus AAI is subaltern to AAA, EAO to EAE, and so on with the rest.

§ 629. The remaining 19 combinations of mood and figure, which are loosely called 'moods,' though in strictness they should be called 'figured moods,' are generally spoken of under the names supplied by the following mnemonics—

Barbara, Celarent, Darii, Ferioque prioris;
Cesare, Camestres, Festino, Baroko secundæ;
Tertia Darapti, Disamis, Datisi, Felapton,
Bokardo, Ferison habet; Quarta insuper addit

Bramantip, Camenes, Dimaris, Fesapo, Fresison:

Quinque Subalterni, totidem Generalibus orti,

Nomen habent nullum, nee, si bene colligis, usum.

§ 630. The vowels in these lines indicate the letters of the mood. All the special rules of the four figures can be gathered from an inspection of them. The following points should be specially noted.

The first figure proves any kind of conclusion, and is the only one which can prove A.

The second figure proves only negatives.

The third figure proves only particulars.

The fourth figure proves any conclusion except A.

§ 631. The first figure is called the Perfect, and the rest the Imperfect figures. The claim of the first to be regarded as the perfect figure may be rested on these grounds—

1. It alone conforms directly to the Dictum de Omni et Nullo.

2. It suffices to prove every kind of conclusion, and is the only figure in which a universal affirmative proposition can be established.

3. It is only in a mood of this figure that the major, middle and minor terms are to be found standing in their relative order of extension.

§ 632. The reason why a universal affirmative, which is of course infinitely the most important form of proposition, can only be proved in the first figure may be seen as follows.

Proof that A can only be established in figure I.

An A conclusion necessitates both premisses being A propositions (by Rule 7). But the minor term is distributed in the conclusion, as being the subject of an A proposition, and must therefore be distributed in the minor premiss, in order to which it must be the subject. Therefore the middle term must be the predicate and is consequently undistributed. In order therefore that the middle term may be distributed, it must be subject in the major premiss, since that also is an A proposition. But when the middle term is subject in the major and predicate in the minor premiss, we have what is called the first figure.

CHAPTER XV.

Of the Special Canons of the Four Figures.

§ 633. So far we have given only a negative test of legitimacy, having shown what moods are not invalidated by running counter to any of the special rules of the four figures. We will now lay down special canons for the four figures, conformity to which will serve as a positive test of the validity of a given mood in a given figure. The special canon of the first figure—will of course be practically equivalent to the Dictum de Omni et Nullo. All of them will be expressed in terms of extension, for the sake of perspicuity.

Special Canons of the Four Figures.

FIGURE 1.

§ 634. CANON. If one term wholly includes or excludes another, which wholly or partly includes a third, the first term wholly or partly includes or excludes the third.

Here four cases arise—

[Illustration]

(1) Total inclusion (Barbara).

All B is A.
All C is B.
∴. All C is A.
[Illustration]
(2) Partial inclusion (Darii).
All B is A.
Some C is B.
∴. Some C is A.
[Illustration]
(3) Total exclusion (Celarent).
No B is A.
All C is B.
∴. No C is A.
[Illustration]
(4) Partial exclusion (Ferio).
No B is A.
Some C is B.
∴. Some C is not A.
FIGURE II.

§ 635. CANON. If one term is excluded from another, which wholly or partly includes a third, or is included in another from which a third is wholly or partly excluded, the first is excluded from the whole or part of the third.

Here we have four cases, all of exclusion—

(1) Total exclusion on the ground of inclusion in an excluded term (Cesare).
[Illustration]
No A is B.
All C is B.
∴. No C is A.

(2) Partial exclusion on the ground of a similar partial inclusion (Festino).
[Illustration]
No A is B.
Some C is B.
∴. Some C is not A.

(3) Total exclusion on the ground of exclusion from an including term (Camestres).
[Illustration]
All A is B.
No C is B.
∴. No C is A.

(4) Partial exclusion on the ground of a similar partial exclusion (Baroko).
[Illustration]
All A is B.
Some C is not B.
∴. Some C is not A.
FIGURE III.

§ 636. CANON. If two terms include another term in common, or if the first includes the whole and the second a part of the same term, or vice versâ, the first of these two terms partly includes the second; and if the first is excluded from the whole of a term which is wholly or in part included in the second, or is excluded from part of a term which is wholly included in the second, the first is excluded from part of the second.

Here it is evident from the statement that six cases arise—

(1) Total inclusion of the same term in two others (Darapti).

[Illustration]

All B is A. All B is C. .'. some C is A.

(2) Total inclusion in the first and partial inclusion in the second (Datisi).

[Illustration]

All B is A. Some B is C. .'. some C is A.

(3) Partial inclusion in the first and total inclusion in the second (Disamis).

[Illustration]

Some B is A. All B is C. .'. some C is A.

(4) Total exclusion of the first from a term which is wholly included in the second (Felapton).

[Illustration]

No B is A. All B is C. .'. some C is not A.

(5) Total exclusion of the first from a term which is partly included in the second (Ferison).

[Illustration]

No B is A. Some B is C. .'. some C is not A.

(6) Exclusion of the first from part of a term which is wholly included in the second (Bokardo).

[Illustration]

Some B is not A.

All B is C.

.'. Some C is not A.

FIGURE IV.

§ 637. CANON. If one term is wholly or partly included in another which is wholly included in or excluded from a third, the third term wholly or partly includes the first, or, in the case of total inclusion, is wholly excluded from it; and if a term is excluded from another which is wholly or partly included in a third, the third is partly excluded from the first.

Here we have five cases—

(1) Of the inclusion of a whole term (Bramsntip).

[Illustration]

All A is B.

All B is C.

.'. Some C is (all) A.

(2) Of the inclusion of part of a term (DIMARIS).

[Illustration]

Some A is B.

All B is C.

.'. Some C is (some) A,

(3) Of the exclusion of a whole term (Camenes).

[Illustration]

All A is B.

No B is C.

.'. No C is A.

(4) Partial exclusion on the ground of including the whole of an excluded term (Fesapo).

[Illustration]

81

No A is B.

All B is C.

.'. Some C is not A.

(5) Partial exclusion on the ground of including part of an excluded term (Fresison).

[Illustration]

No A is B.

Some B is C.

.'. Some C is not A.

§ 638. It is evident from the diagrams that in the subaltern moods the conclusion is not drawn directly from the premises, but is an immediate inference from the natural conclusion. Take for instance AAI in the first figure. The natural conclusion from these premises is that the minor term C is wholly contained in the major term A. But instead of drawing this conclusion we go on to infer that something which is contained in C, namely some C, is contained in A.

[Illustration]

All B is A. All C is B. .'. all C is A. .'. some C is A.

Similarly in EAO in figure 1, instead of arguing that the whole of C is excluded from A, we draw a conclusion which really involves a further inference, namely that part of C is excluded from A.

[Illustration]

No B is A. All C is B. .'. no C is A. .'. some C is not A.

§ 639. The reason why the canons have been expressed in so cumbrous a form is to render the validity of all the moods in each figure at once apparent from the statement. For purposes of general convenience they admit of a much more compendious mode of expression.

§ 640. The canon of the first figure is known as the Dictum de Omni et Nullo—

What is true (distributively) of a whole term is true of all that it includes.

§ 641. The canon of the second figure is known as the Dictum de Diverse—

If one term is contained in, and another excluded from a third term, they are mutually excluded.

§ 642. The canon of the third figure is known as the Dictum de Exemplo et de Excepto—

Two terms which contain a common part partly agree, or, if one contains a part which the other does not, they partly differ.

§ 643. The canon of the fourth figure has had no name assigned to it, and does not seem to admit of any simple expression. Another mode of formulating it is as follows:—

Whatever is affirmed of a whole term may have partially affirmed of it whatever is included in that term (Bramantip, Dimaris), and partially denied of it whatever is excluded (Fesapo); whatever is affirmed of part of a term may have partially denied of it whatever is wholly excluded from that term (Fresison); and whatever is denied of a whole term may have wholly denied of it whatever is wholly included in that term (Camenes).

§ 644. From the point of view of intension the canons of the first three figures may be expressed as follows.

§ 645. Canon of the first figure. Dictum de Omni et Nullo—

An attribute of an attribute of anything is an attribute of the thing itself.

§ 646. Canon of the second figure. Dictum de Diverso—

If a subject has an attribute which a class has not, or vice versa, the subject does not belong to the class.

§ 647. Canon of the third figure.

1. Dictum de Exemplo—
> If a certain attribute can be affirmed of any portion of the members of a class, it is not incompatible with the distinctive attributes of that class.

2. Dictum de Excepto—
> If a certain attribute can be denied of any portion of the members of a class, it is not inseparable from the distinctive attributes of that class.

CHAPTER XVI.

Of the Special Uses of the Four Figures.

§ 648. The first figure is useful for proving the properties of a thing.

§ 649. The second figure is useful for proving distinctions between things.

§ 650. The third figure is useful for proving instances or exceptions.

§ 651. The fourth figure is useful for proving the species of a genus.

FIGURE 1.

§ 652.

B is or is not A.

C is B.

∴ C is or is not A.

We prove that C has or has not the property A by predicating of it B, which we know to possess or not to possess that property.

Luminous objects are material.

Comets are luminous.

∴ Comets are material.

No moths are butterflies.

The Death's head is a moth.

∴ The Death's head is not a butterfly.

FIGURE II.

§ 653.

A is B. A is not B.

C is not B. C is B.

∴ C is not A. ∴ C is not A.

We establish the distinction between C and A by showing that A has an attribute which C is devoid of, or is devoid of an attribute which C has.

All fishes are cold-blooded.

A whale is not cold-blooded.

∴ A whale is not a fish.

No fishes give milk.

A whale gives milk.

∴ A whale is not a fish.

FIGURE III.

§ 654.

B is A. B is not A.

B is C. B is C.

∴ Some C is A. ∴ Some C is not A.

We produce instances of C being A by showing that C and A meet, at all events partially, in B. Thus if we wish to produce an instance of the compatibility of great learning with original powers of thought, we might say

Sir William Hamilton was an original thinker.
Sir William Hamilton was a man of great learning.
.'. Some men of great learning are original thinkers.

Or we might urge an exception to the supposed rule about Scotchmen being deficient in humour under the same figure, thus—

Sir Walter Scott was not deficient in humour.
Sir Walter Scott was a Scotchman.
.'. Some Scotchmen are not deficient in humour.

FIGURE IV.

§ 655.

All A is B, No A is B.
All B is C. All B is C.
.'. Some C is A .'. Some C is not A.

We show here that A is or is not a species of C by showing that A falls, or does not fall, under the class B, which itself falls under C. Thus—

All whales are mammals.
All mammals are warm-blooded.
.'. Some warm-blooded animals are whales.
No whales are fishes.
All fishes are cold-blooded.
.'. Some cold-blooded animals are not whales.

CHAPTER XVII.

Of the Syllogism with three figures.

§ 656. It will be remembered that in beginning to treat of figure (§ 565) we pointed out that there were either four or three ligures possible according as the conclusion was assumed to be known or not. For, if the conclusion be not known, we cannot distinguish between the major and the minor term, nor, consequently, between one premiss and another. On this view the first and the fourth figures are the same, being that arrangement of the syllogism in which the middle term occupies a different position in one premiss from what it does in the other. We will now proceed to constitute the legitimate moods and figures of the syllogism irrespective of the conclusion.

§ 657. When the conclusion is set out of sight, the number of possible moods is the same as the number of combinations that can be made of the four things, A, E, I, O, taken two together, without restriction as to repetition. These are the following 16:—

AA EA IA OA AE -EE- IE -OE- AI EI -II- -OI- AO -EO- -IO- -OO-

of which seven may be neglected as violating the general rules of the syllogism, thus leaving us with nine valid moods—

AA. AE. AI. AO. EA. EI. IA. IE. OA.

§ 658. We will now put these nine moods successively into the three figures. By so doing it will become apparent how far they are valid in each.

§ 659. Let it be premised that

when the extreme in the premiss that stands first is predicate in the conclusion, we are said to have a Direct Mood;

when the extreme in the premiss that stands second is predicate in the conclusion, we are said to have an Indirect Mood.

§ 660. FIGURE 1.

Mood AA.

All B is A.

All C is B.

∴. All C is A, or Some A is C, (Barbara & Bramantip).

Mood AE.

All B is A.

No C is B.

∴. Illicit Process, or Some A is not C, (Fesapo).

Mood AI.

All B is A.

Some C is B.

∴. Some C is A, or Some A is C. (Darii & Disamis).

Mood AO.

All B is A.

Some C is not B.

∴. Illicit Process, (Ferio).

Mood EA.

No B is A.

All C is B.

∴. No C is A, or No A is C, (Celarent & Camenes).

Mood EI.

No B is A.

Some C is B.

∴. Some C is not A, or Illicit Process.

Mood IA.

Some B is A.

All C is B.

∴. Undistributed Middle.

Mood IE.

Some B is C. Some B is not A.

No A is B. All C is B.

∴. Illicit Process, or Some C is not A, (Fresison).

Mood OA.

Some B is not A.

All C is B.

∴. Undistributed Middle.

§ 661. Thus we are left with six valid moods, which yield four direct conclusions and five indirect ones, corresponding to the four moods of the original first figure and the five moods of the original fourth, which appear now as indirect moods of the first figure.

§ 662. But why, it maybe asked, should not the moods of the first figure equally well be regarded as indirect moods of the fourth? For this reason-that all the moods of the fourth figure can be elicited out of premisses in which the terms stand in the order of the first, whereas the converse is not the case. If, while retaining the quantity and quality of the above premisses, i. e. the mood, we were in each case to transpose the terms, we should find that we were left with five valid moods instead of six, since AI in the reverse order of the terms involves undistributed middle; and, though we should have Celarent indirect to Camenes, and Darii to Dimaris, we should never arrive at the conclusion of Barbara or

85

have anything exactly equivalent to Ferio. In place of Barbara, Bramantip would yield as an indirect mood only the subaltern AAI in the first figure. Both Fesapo and Fresison would result in an illicit process, if we attempted to extract the conclusion of Ferio from them as an indirect mood. The nearest approach we could make to Ferio would be the mood EAO in the first figure, which may be elicited indirectly from the premisses of CAMENES, being subaltern to CELARENT. For these reasons the moods of the fourth figure are rightly to be regarded as indirect moods of the first, and not vice versâ.

$663. FIGURE II.

Mood AA.
All A is B.
All C is B.
.'. Undistributed Middle.

Mood AE.
All A is B.
No C is B.
.'. No C is A, or No A is C, (Camestres & Cesare).

Mood AI.
All A is B.
Some C is B.
.'. Undistributed Middle.

Mood AO.
All A is B.
Some C is not B.
.'. Some C is not A, (Baroko), or Illicit Process.

Mood EA.
No A is B.
All C is B.
.'. No C is A, or No A is C, (Cesare & Carnestres).

Mood EI
No A is B.
Some C is B.
.'. Some C is not A, (Festino), or Illicit Process.

Mood IA.
Some A is B.
All C is B.
.'. Undistributed Middle.

Mood IE.
Some A is B.
No C is B.
.'. Illicit Process, or Some A is not C, (Festino).

Mood OA.
Some A is not B.
All C is B.
.'. Illicit Process, or Some A is not C, (Baroko).

§ 664. Here again we have six valid moods, which yield four direct conclusions corresponding to Cesare, CARNESTRES, FESTINO and BAROKO. The same four are repeated in the indirect moods.

§ 665. FIGURE III.

Mood AA.
All B is A.
All B is C.
.'. Some C is A, or Some A is C, (Darapti).

86

Mood AE.
All B is A.
No B is C.
∴ Illicit Process, or Some A is not C, (Felapton).
Mood AI.
All B is A,
Some B is C.
∴ Some C is A, or Some A is C, (Datisi & Disamis).
Mood AO.
All B is A.
Some B is not C.
∴ Illicit Process, Or Some A is not C, (Bokardo).
Mood EA.
No B is A.
All B is C.
∴ Some C is not A, (Felapton), or Illicit Process.
Mood EI.
No B is A.
Some B is C.
∴ Some C is not A, (Ferison), or Illicit Process.
Mood IA.
Some B is A.
All B is C.
∴ Some C is A, Or Some A is C, (Disamis & Datisi).
Mood IE.
Some B is A.
No B is C.
∴ Illicit Process, or Some A is not C, (Ferison).
Mood QA.
Some B is not A.
All B is C.
∴ Some C is not A, (Bokardo), or Illicit Process.

§ 666. In this figure every mood is valid, either directly or indirectly. We have six direct moods, answering to Darapti, Disamis, Datisi, Felapton, Bokardo and Ferison, which are simply repeated by the indirect moods, except in the case of Darapti, which yields a conclusion not provided for in the mnemonic lines. Darapti, though going under one name, has as much right to be considered two moods as Disamis and Datisi.

CHAPTER XVIII.

Of Reduction.

§ 667. We revert now to the standpoint of the old logicians, who regarded the Dictum de Omni et Nullo as the principle of all syllogistic reasoning. From this point of view the essence of mediate inference consists in showing that a special case, or class of cases, comes under a general rule. But a great deal of our ordinary reasoning does not conform to this type. It was therefore judged necessary to show that it might by a little manipulation be brought into conformity with it. This process is called Reduction.

§ 668. Reduction is of two kinds—

(1) Direct or Ostensive.

(2) Indirect or Ad Impossibile.

§ 669. The problem of direct, or ostensive, reduction is this—

Given any mood in one of the imperfect figures (II, III and IV) how to alter the form of the premisses so as to arrive at the same conclusion in the perfect figure, or at one from which it can be immediately inferred. The alteration of the premisses is effected by means of immediate inference and, where necessary, of transposition.

§ 670. The problem of indirect reduction, or reductio (per deductionem) ad impossibile, is this—Given any mood in one of the imperfect figures, to show by means of a syllogism in the perfect figure that its conclusion cannot be false.

§ 671. The object of reduction is to extend the sanction of the Dictum de Omni et Nullo to the imperfect figures, which do not obviously conform to it.

§ 672. The mood required to be reduced is called the Reducend; that to which it conforms, when reduced, is called the Reduct.

Direct or Ostensive Reduction.

§ 673. In the ordinary form of direct reduction, the only kind of immediate inference employed is conversion, either simple or by limitation; but the aid of permutation and of conversion by negation and by contraposition may also be resorted to.

§ 674. There are two moods, Baroko and Bokardo, which cannot be reduced ostensively except by the employment of some of the means last mentioned. Accordingly, before the introduction of permutation into the scheme of logic, it was necessary to have recourse to some other expedient, in order to demonstrate the validity of these two moods. Indirect reduction was therefore devised with a special view to the requirements of Baroko and Bokardo: but the method, as will be seen, is equally applicable to all the moods of the imperfect figures.

§ 675. The mnemonic lines, 'Barbara, Celarent, etc., provide complete directions for the ostensive reduction of all the moods of the second, third, and fourth figures to the first, with the exception of Baroko and Bokardo. The application of them is a mere mechanical trick, which will best be learned by seeing the process performed.

§ 676. Let it be understood that the initial consonant of each name of a figured mood indicates that the reduct will be that mood which begins with the same letter. Thus the B of Bramantip indicates that Bramantip, when reduced, will become Barbara.

§ 677. Where m appears in the name of a reducend, me shall have to take as major that premiss which before was minor, and vice versa-in other words, to transpose the premisses, m stands for mutatio or metathesis.

§ 678. s, when it follows one of the premisses of a reducend, indicates that the premiss in question must be simply converted; when it follows the conclusion, as in Disamis, it indicates that the conclusion arrived at in the first figure is not identical in form with the original conclusion, but capable of being inferred from it by simple conversion. Hence s in the middle of a name indicates something to be done to the original premiss, while s at the end indicates something to be done to the new conclusion.

§ 679. P indicates conversion per accidens, and what has just been said of s applies, mutatis mutandis, to p.

§ 680. k may be taken for the present to indicate that Baroko and Bokardo cannot be reduced ostensively.

§ 681. FIGURE II.

Cesare. \ / Celarent.

No A is B. \ = / No B is A.

All C is B. / \ All C is B.

∴. No C is A. / \ ∴. No C is A.

Camestres. \ / Celarent.
All A is B. \ = / No B is C.
No C is B. / \ All A is B.
.'. No C is A. / \ .'. No A is C.
.'. No C is A.
Festino. Ferio.
No A is B. \ / No B is A.
Some C is B. | = | Some C is B.
.'. Some C is not A./ \ .'. Some C is not A.
[Baroko]
§ 682. FIGURE III.
Darapti. \ / Darii.
All B is A. \ = / All B is A.
All B is C. / \ Some C is B.
.'. Some C is A. / \ Some C is A.
Disamis. \ / Darii.
Some B is A. \ = / All B is C.
All B is C. / \ Some A is B.
.'. Some C is A. / \ .'. Some A is C.
.'. Some C is A.
Datisi. \ / Darii.
All B is A. \ = / All B is A.
Some B is C. / \ Some C is B.
.'. Some C is A. / \ .'. Some C is A.
Felapton. \ / Ferio.
No B is A. \ = / No B is A.
All B is C. / \ Some C is B.
.'. Some C is not-A. / \ .'. Some C is not-A.
[Bokardo].
Ferison. \ / Ferio.
No B is A. \ = / No B is A.
Some B is C. / \ Some C is B
.'. Some C is not A. / \ .'. Some C is not A.
§ 683. FIGURE IV.
Bramantip. \ / Barbara.
All A is B. \ = / All B is C.
All B is C. / \ All A is B.
.. Some C is A. / \ .. All A is C.
.'. Some C is A.
Camenes Celarent
All A is B \ / No B is C.
No B is C. | = | All A is B.
.. No C is A./ \ .'. No A is C.
.'. No C is A.
Dimaris. Darii.
Some A is B. \ / All B is C.
All B is C. | = | Some A is B.
.'. Some C is A./ \ .'. Some A is C.
.'. Some C is A.
Fesapo. Ferio.
No A is B. \ / No B is A.

All B is C. | = | Some C is B.
∴ Some C is not A./ \ ∴ Some C is not A.
   Fresison. Ferio.
No A is B. \ / No B is A.
Some B is C. | = | Some C is B.
∴ Some C is not A./ \ ∴ Some C is not A.

§ 684. The reason why Baroko and Bokardo cannot be reduced ostensibly by the aid of mere conversion becomes plain on an inspection of them. In both it is necessary, if we are to obtain the first figure, that the position of the middle term should be changed in one premiss. But the premisses of both consist of A and 0 propositions, of which A admits only of conversion by limitation, the effect of which would be to produce two particular premisses, while 0 does not admit of conversion at all,

It is clear then that the 0 proposition must cease to be 0 before we can get any further. Here permutation comes to our aid; while conversion by negation enables us to convert the A proposition, without loss of quantity, and to elicit the precise conclusion we require out of the reduct of Boltardo.

   (Baroko) Fanoao. Ferio.
All A is B. \ / No not-B is A.
Some C is not-B. | = | Some C is not-B.
∴ Some C is not-A./ \ ∴ Some C is not-A.
   (Bokardo) Donamon. Darii.
Some B is not-A. \ / All B is C.
All B is C. | = | Some not-A is B
∴ Some C is not-A./ \ ∴ Some not-A is C.
          ∴ Some C is not-A.

§ 685. In the new symbols, Fanoao and Donamon, [pi] has been adopted as a symbol for permutation; n signifies conversion by negation. In Donamon the first n stands for a process which resolves itself into permutation followed by simple conversion, the second for one which resolves itself into simple conversion followed by permutation, according to the extended meaning which we have given to the term 'conversion by negation.' If it be thought desirable to distinguish these two processes, the ugly symbol Do[pi]samos[pi] may be adopted in place of Donamon.

§ 686. The foregoing method, which may be called Reduction by Negation, is no less applicable to the other moods of the second figure than to Baroko. The symbols which result from providing for its application would make the second of the mnemonic lines run thus—

Benare[pi], Cane[pi]e, Denilo[pi], Fano[pi]o secundae.

§ 687. The only other combination of mood and figure in which it will be found available is Camenes, whose name it changes to Canene.

   § 688.
   (Cesare) Benarea. Barbara.
No A is B. \ / All B is not-A.
All C is B. | = | All C is B.
∴ No C is A. / \ ∴ All C is not-A.
          ∴ No C is A.
   (Camestres) Cane[pi]e. Celarent.
All A is B. \ / No not-B is A.
No C is B. | = | All C is not-B.
∴ No C is A. / \ ∴ No C is A.
   (Festino) Denilo[pi]. Darii.
No A is B. \ / All B is not-A.
Some C is B. | = | Some C is B.

.'. Some C is not A./ \ .'. Some C is not-A.

.'. Some C is not A.

(Camenes) Canene. Celarent.

All A is B. \ / No not-B is A.

No B is C. | = | All C is not-B.

.'. No C is A. / \ .'. No C is A.

§ 689. The following will serve as a concrete instance of Cane[pi]e reduced to the first figure.

All things of which we have a perfect idea are perceptions.

A substance is not a perception.

.'. A substance is not a thing of which we have a perfect idea.

When brought into Celarent this becomes—

No not-perception is a thing of which we have a perfect idea.

A substance is a not-perception.

.'. No substance is a thing of which we have a perfect idea.

§ 690. We may also bring it, if we please, into Barbara, by permuting the major premiss once more, so as to obtain the contrapositive of the original—

All not-perceptions are things of which we have an imperfect idea.

All substances are not-perceptions.

.'. All substances are things of which we have an imperfect idea.

Indirect Reduction.

§ 691. We will apply this method to Baroko.

All A is B. All fishes are oviparous.

Some C is not B. Some marine animals are not oviparous.

.'. Some C is not A. .'. Some marine animals are not fishes.

§ 692. The reasoning in such a syllogism is evidently conclusive: but it does not conform, as it stands, to the first figure, nor (permutation apart) can its premisses be twisted into conformity with it. But though we cannot prove the conclusion true in the first figure, we can employ that figure to prove that it cannot be false, by showing that the supposition of its falsity would involve a contradiction of one of the original premisses, which are true ex hypothesi.

§ 693. If possible, let the conclusion 'Some C is not A' be false. Then its contradictory 'All C is A' must be true. Combining this as minor with the original major, we obtain premisses in the first figure,

All A is B, All fishes are oviparous,

All C is A, All marine animals are fishes,

which lead to the conclusion

All C is B, All marine animals are oviparous.

But this conclusion conflicts with the original minor, 'Some C is not B,' being its contradictory. But the original minor is ex hypothesi true. Therefore the new conclusion is false. Therefore it must either be wrongly drawn or else one or both of its premisses must be false. But it is not wrongly drawn; since it is drawn in the first figure, to which the Dictum de Omni et Nullo applies. Therefore the fault must lie in the premisses. But the major premiss, being the same with that of the original syllogism, is ex hypothesi true. Therefore the minor premiss, 'All C is A,' is false. But this being false, its contradictory must be true. Now its contradictory is the original conclusion, 'Some C is not A,' which is therefore proved to be true, since it cannot be false.

§ 694. It is convenient to represent the two syllogisms in juxtaposition thus—

Baroko. Barbara.

All A is B. All A is B.

Some C is not B. \/ All C is A.

.'. Some C is not A. /\ All C is B.

§ 695. The lines indicate the propositions which conflict with one another. The initial consonant of the names Baroko and Eokardo indicates that the indirect reduct will be Barbara. The k indicates that the O proposition, which it follows, is to be dropped out in the new syllogism, and its place supplied by the contradictory of the old conclusion.

§ 696. In Bokardo the two syllogisms will stand thus—

Bokardo. Barbara.
Some B is not A. \ / All C is A.
All B is C. X All B is C.
∴ Some C is not A./ \ ∴ All B is A.

§ 697. The method of indirect reduction, though invented with a special view to Baroko and Bokardo, is applicable to all the moods of the imperfect figures. The following modification of the mnemonic lines contains directions for performing the process in every case:—Barbara, Celarent, Darii, Ferioque prioris; Felake, Dareke, Celiko, Baroko secundae; Tertia Cakaci, Cikari, Fakini, Bekaco, Bokardo, Dekilon habet; quarta insuper addit Cakapi, Daseke, Cikasi, Cepako, Cesïkon.

§ 698. The c which appears in two moods of the third figure, Cakaci and Bekaco, signifies that the new conclusion is the contrary, instead of, as usual, the contradictory of the discarded premiss.

§ 699. The letters s and p, which appear only in the fourth figure, signify that the new conclusion does not conflict directly with the discarded premiss, but with its converse, either simple or per accidens, as the case may be.

§ 700. l, n and r are meaningless, as in the original lines.

CHAPTER XIX.

Of Immediate Inference as applied to Complex Propositions.

§ 701. So far we have treated of inference, or reasoning, whether mediate or immediate, solely as applied to simple propositions. But it will be remembered that we divided propositions into simple and complex. I t becomes incumbent upon us therefore to consider the laws of inference as applied to complex propositions. Inasmuch however as every complex proposition is reducible to a simple one, it is evident that the same laws of inference must apply to both.

§ 702. We must first make good this initial statement as to the essential identity underlying the difference of form between simple and complex propositions.

§ 703. Complex propositions are either Conjunctive or Disjunctive (§ 214).

§ 704. Conjunctive propositions may assume any of the four forms, A, E, I, O, as follows—

(A) If A is B, C is always D.
(E) If A is B, C is never D.
(I) If A is B, C is sometimes D.
(O) If A is B, C is sometimes not D.

§ 705. These admit of being read in the form of simple propositions, thus—

(A) If A is B, C is always D = All cases of A being B are cases of C being D. (Every AB is a CD.)

(E) If A is B, C is never D = No cases of A being B are cases of C being D. (No AB is a CD.)

(I) If A is B, C is sometimes D = Some cases of A being B are cases of C being D. (Some AB's are CD's.)

(O) If A is B, C is sometimes not D = Some cases of A being B are not cases of C being D. (Some AB's are not CD's.)

§ 706. Or, to take concrete examples,

(A) If kings are ambitious, their subjects always suffer.

= All cases of ambitious kings are cases of subjects suffering.

(E) If the wind is in the south, the river never freezes.

= No cases of wind in the south are cases of the river freezing.

(I) If a man plays recklessly, the luck sometimes goes against him.

= Some cases of reckless playing are cases of going against one.

(O) If a novel has merit, the public sometimes do not buy it.

= Some cases of novels with merit are not cases of the public buying.

§ 707. We have seen already that the disjunctive differs from the conjunctive proposition in this, that in the conjunctive the truth of the antecedent involves the truth of the consequent, whereas in the disjunctive the falsity of the antecedent involves the truth of the consequent. The disjunctive proposition therefore

Either A is B or C is D

may be reduced to a conjunctive

If A is not B, C is D,

and so to a simple proposition with a negative term for subject.

All cases of A not being B are cases of C being D.

(Every not-AB is a CD.)

§ 708. It is true that the disjunctive proposition, more than any other form, except U, seems to convey two statements in one breath. Yet it ought not, any more than the E proposition, to be regarded as conveying both with equal directness. The proposition 'No A is B' is not considered to assert directly, but only implicitly, that 'No B is A.' In the same way the form 'Either A is B or C is D' ought to be interpreted as meaning directly no more than this, 'If A is not B, C is D.' It asserts indeed by implication also that 'If C is not D, A is B.' But this is an immediate inference, being, as we shall presently see, the contrapositive of the original. When we say 'So and so is either a knave or a fool,' what we are directly asserting is that, if he be not found to be a knave, he will be found to be a fool. By implication we make the further statement that, if he be not cleared of folly, he will stand condemned of knavery. This inference is so immediate that it seems indistinguishable from the former proposition: but since the two members of a complex proposition play the part of subject and predicate, to say that the two statements are identical would amount to asserting that the same proposition can have two subjects and two predicates. From this point of view it becomes clear that there is no difference but one of expression between the disjunctive and the conjunctive proposition. The disjunctive is merely a peculiar way of stating a conjunctive proposition with a negative antecedent.

§ 709. Conversion of Complex Propositions.

A / If A is B, C is always D.

\ .'. If C is D, A is sometimes B.

E / If A is B, C is never D.

\ .'. If C is D, A is never B.

I / If A is S, C is sometimes D.

\ .'. If C is D, A is sometimes B.

§ 710. Exactly the same rules of conversion apply to conjunctive as to simple propositions.

§ 711. A can only be converted per accidens, as above.

The original proposition

'If A is B, C is always D'
is equivalent to the simple proposition
'All cases of A being B are cases of C being D.'
This, when converted, becomes
'Some cases of C being D are cases of A being B,'
which, when thrown back into the conjunctive form, becomes
'If C is D, A is sometimes B.'

§ 712. This expression must not be misunderstood as though it contained any reference to actual existence. The meaning might be better conveyed by the form
'If C is D, A may be B.'
But it is perhaps as well to retain the other, as it serves to emphasize the fact that formal logic is concerned only with the connection of ideas.

§ 713. A concrete instance will render the point under discussion clearer. The example we took before of an A proposition in the conjunctive form—
'If kings are ambitious, their subjects always suffer'
may be converted into
'If subjects suffer, it may be that their kings are ambitious,'
i.e. among the possible causes of suffering on the part of subjects is to be found the ambition of their rulers, even if every actual case should be referred to some other cause. It is in this sense only that the inference is a necessary one. But then this is the only sense which formal logic is competent to recognise. To judge of conformity to fact is no part of its province. From 'Every AB is a CD' it follows that ' Some CD's are AB's' with exactly the same necessity as that with which 'Some B is A' follows from 'All A is B.' In the latter case also neither proposition may at all conform to fact. From 'All centaurs are animals' it follows necessarily that 'Some animals are centaurs': but as a matter of fact this is not true at all.

§ 714. The E and the I proposition may be converted simply, as above.

§ 715. O cannot be converted at all. From the proposition
'If a man runs a race, he sometimes does not win it,'
it certainly does not follow that
'If a man wins a race, he sometimes does not run it.'

§ 716. There is a common but erroneous notion that all conditional propositions are to be regarded as affirmative. Thus it has been asserted that, even when we say that 'If the night becomes cloudy, there will be no dew,' the proposition is not to be regarded as negative, on the ground that what we affirm is a relation between the cloudiness of night and the absence of dew. This is a possible, but wholly unnecessary, mode of regarding the proposition. It is precisely on a par with Hobbes's theory of the copula in a simple proposition being always affirmative. It is true that it may always be so represented at the cost of employing a negative term; and the same is the case here.

§ 717. There is no way of converting a disjunctive proposition except by reducing it to the conjunctive form.

§ 718. *Permutation of Complex Propositions.*
  (A) If A is B, C is always D.
  ∴. If A is B, C is never not-D. (E)
  (E) If A is B, C is never D.
  ∴. If A is B, C is always not-D. (A)
  (I) If A is B, C is sometimes D.
  ∴. If A is B, C is sometimes not not-D. (O)
  (O) If A is B, C is sometimes not D.
  ∴. If A is B, C is sometimes not-D. (I)
§ 719.

(A) If a mother loves her children, she is always kind to them.

∴ If a mother loves her children, she is never unkind to them. (E)

(E) If a man tells lies, his friends never trust him.

∴ If a man tells lies, his friends always distrust him. (A)

(I) If strangers are confident, savage dogs are sometimes friendly.

∴ If strangers are confident, savage dogs are sometimes not unfriendly. (O)

(O) If a measure is good, its author is sometimes not popular.

∴ If a measure is good, its author is sometimes unpopular. (I)

§ 720. The disjunctive proposition may be permuted as it stands without being reduced to the conjunctive form.

Either A is B or C is D.

∴ Either A is B or C is not not-D.

Either a sinner must repent or he will be damned.

∴ Either a sinner must repent or he will not be saved.

§ 721. *Conversion by Negation of Complex Propositions.*

(A) If A is B, C is always D.

∴ If C is not-D, A is never B. (E)

(E) If A is B, C is never D.

∴ If C is D, A is always not-B. (A)

(I) If A is B, C is sometimes D.

∴ If C is D, A is sometimes not not-B. (O)

(O) If A is B, C is sometimes not D.

∴ If C is not-D, A is sometimes B. (I)

(E per acc.) If A is B, C is never D.

∴ If C is not-D, A is sometimes B. (I)

(A per ace.) If A is B, C is always D.

∴ If C is D, A is sometimes not not-D. (O)

§ 722.

(A) If a man is a smoker, he always drinks.

∴ If a man is a total abstainer, he never smokes. (E)

(E) If a man merely does his duty, no one ever thanks him.

∴ If people thank a man, he has always done more than his duty. (A)

(I) If a statesman is patriotic, he sometimes adheres to a party.

∴ If a statesman adheres to a party, he is sometimes not unpatriotic. (O)

(O) If a book has merit, it sometimes does not sell.

∴ If a book fails to sell, it sometimes has merit. (I)

(E per acc.) If the wind is high, rain never falls.

∴ If rain falls, the wind is sometimes high. (I)

(A per acc.) If a thing is common, it is always cheap.

∴ If a thing is cheap, it is sometimes not uncommon. (O)

§ 723. When applied to disjunctive propositions, the distinctive features of conversion by negation are still discernible. In each of the following forms of inference the converse differs in quality from the convertend and has the contradictory of one of the original terms (§ 515).

§ 724.

(A) Either A is B or C is always D.

∴ Either C is D or A is never not-B. (E)

(E) Either A is B or C is never D.

∴ Either C is not-D or A is always B. (A)

(I) Either A is B or C is sometimes D.

∴ Either C is not-D or A is sometimes not B. (O)

(O) Either A is B or C is sometimes not D.

∴ Either C is D or A is sometimes not-B. (I)

§ 725.

(A) Either miracles are possible or every ancient historian is untrustworthy.

∴ Either ancient historians are untrustworthy or miracles are not impossible. (E)

(E) Either the tide must turn or the vessel can not make the port.

∴ Either the vessel cannot make the port or the tide must turn. (A)

(1) Either he aims too high or the cartridges are sometimes bad.

∴ Either the cartridges are not bad or he sometimes does not aim too high. (0)

(O) Either care must be taken or telegrams will sometimes not be correct.

∴ Either telegrams are correct or carelessness is sometimes shown. (1)

§ 726. In the above examples the converse of E looks as if it had undergone no change but the mere transposition of the alternative. This appearance arises from mentally reading the E as an A proposition: but, if it were so taken, the result would be its contrapositive, and not its converse by negation.

§ 727. The converse of I is a little difficult to grasp. It becomes easier if we reduce it to the equivalent conjunctive—

'If the cartridges are bad, he sometimes does not aim too high.'

Here, as elsewhere, 'sometimes' must not be taken to mean more than 'it may be that.'

§ 728. *Conversion by Contraposition of Complex Propositions.*

As applied to conjunctive propositions conversion by contraposition assumes the following forms—

(A) If A is B, C is always D.

∴ If C is not-D, A is always not-B.

(O) If A is B, C is sometimes not D.

∴ If C is not-D, A is sometimes not not-B.

(A) If a man is honest, he is always truthful.

∴ If a man is untruthful, he is always dishonest.

(O) If a man is hasty, he is sometimes not malevolent.

∴ If a man is benevolent, he is sometimes not unhasty.

§ 729. As applied to disjunctive propositions conversion by contraposition consists simply in transposing the two alternatives.

(A) Either A is B or C is D.

∴ Either C is D or A is B.

For, when reduced to the conjunctive shape, the reasoning would run thus—

If A is not B, C is D.

∴ If C is not D, A is B.

which is the same in form as

All not-A is B.

∴ All not-B is A.

Similarly in the case of the O proposition

(O) Either A is B or C is sometimes not D.

∴. Either C is D or A is sometimes not B.

§ 730. On comparing these results with the converse by negation of each of the same propositions, A and 0, the reader will see that they differ from them, as was to be expected, only in being permuted. The validity of the inference may be tested, both here and in the case of conversion by negation, by reducing the disjunctive proposition to the conjunctive, and so to the simple form, then performing the process as in simple propositions, and finally throwing the converse, when so obtained, back into the disjunctive form. We will show in this manner that the above is really the contrapositive of the 0 proposition.

(O) Either A is B or C is sometimes not D.

= If A is not B, C is sometimes not D.

= Some cases of A not being B are not cases of C being D. (Some A is not B.)

= Some cases of C not being D are not cases of A being B. (Some not-B is not not-A.)

= If C is not D, A is sometimes not B.

= Either C is D or A is sometimes not B.

CHAPTER XX.

Of Complex Syllogisms.

§ 731. A Complex Syllogism is one which is composed, in whole or part, of complex propositions.

§ 732. Though there are only two kinds of complex proposition, there are three varieties of complex syllogism. For we may have

(1) a syllogism in which the only kind of complex proposition employed is the conjunctive;

(2) a syllogism in which the only kind of complex proposition employed is the disjunctive;

(3) a syllogism which has one premiss conjunctive and the other disjunctive.

The chief instance of the third kind is that known as the Dilemma.

Syllogism
```
         _____|_____
        | |
Simple Complex
(Categorical) (Conditional)
                    _____|_____
                   | | |
         Conjunctive Disjunctive Dilemma
         (Hypothetical)
```

The Conjunctive Syllogism.

§ 733. The Conjunctive Syllogism has one or both premisses conjunctive propositions: but if only one is conjunctive, the other must be a simple one.

§ 734. Where both premisses are conjunctive, the conclusion will be of the same character; where only one is conjunctive, the conclusion will be a simple proposition.

§ 735. Of these two kinds of conjunctive syllogisms we will first take that which consists throughout of conjunctive propositions.

The Wholly Conjunctive Syllogism.

§ 736. Wholly conjunctive syllogisms do not differ essentially from simple ones, to which they are immediately reducible. They admit of being constructed in every mood and figure, and the moods of the imperfect figures may be brought into the first by following the ordinary rules of reduction. For instance—

Cesare. Celarent.

If A is B, C is never D. \ / If C is D, A is never B.
If E is F, C is always D. | = | If E is F, C is always D.
∴ If E is F, A is never B. / \ ∴ If E is F, A is never B.

If it is day, the stars never shine.\ /If the stars shine, it is never day.
If it is night, the stars always \=/ If it is night, the stars always
shine. / \ shine.
∴ If it is night, it is never day / \∴ If it is night, it is never day.

Disamis. Darii.

If C is D, A is sometimes B. \ / If C is D, E is always F.
If C is D, E is always F. | = | If A is B, C is sometimes D.
If E is F, A is sometimes B. / \ ∴ If A is B, E is sometimes F.
∴ If E is F, A is sometimes B.

If she goes, I sometimes go. \ / If she goes, he always goes,
If she goes, he always goes. | = | If I go, she sometimes goes.
∴ If he goes, I sometimes go. / \ ∴ If I go, he sometimes goes.
∴ If he goes, I sometimes go.

The Partly Conjunctive Syllogism.

§ 737. It is this kind which is usually meant when the Conjunctive or Hypothetical Syllogism is spoken of.

§ 738. Of the two premisses, one conjunctive and one simple, the conjunctive is considered to be the major, and the simple premiss the minor. For the conjunctive premiss lays down a certain relation to hold between two propositions as a matter of theory, which is applied in the minor to a matter of fact.

§ 739. Taking a conjunctive proposition as a major premiss, there are four simple minors possible. For we may either assert or deny the antecedent or the consequent of the conjunctive.

Constructive Mood. Destructive Mood.

(1) If A is B, C is D. (2) If A is B, C is D.
A is B. C is not D.
∴ C is D. ∴ A is not B.

(3) If A is B, C is D. (4) If A is B, C is D.
A is not B. C is D.
No conclusion. No conclusion.

§ 740. When we take as a minor 'A is not B ' (3), it is clear that we can get no conclusion. For to say that C is D whenever A is B gives us no right to deny that C can be D in the absence of that condition. What we have predicated has been merely inclusion of the case AB in the case CD.

[Illustration]

§ 741. Again, when we take as a minor, 'C is D' (4), we can get no universal conclusion. For though A being B is declared to involve as a consequence C being D, yet it is possible for C to be D under other circumstances, or from other causes. Granting the truth of the proposition 'If the sky falls, we shall catch larks,' it by no means follows that there are no other conditions under which this result can be attained.

§ 742. From a consideration of the above four cases we elicit the following

Canon of the Conjunctive Syllogism.

To affirm the antecedent is to affirm the consequent, and to deny the consequent is to deny the antecedent: but from denying the antecedent or affirming the consequent no conclusion follows.

§ 743. There is a case, however, in which we can legitimately deny the antecedent and affirm the consequent of a conjunctive proposition, namely, when the relation predicated between the antecedent and the consequent is not that of inclusion but of coincidence—where in fact the conjunctive proposition conforms to the type u.

For example—

Denial of the Antecedent.

If you repent, then only are you forgiven.

You do not repent.

∴ You are not forgiven.

Affirmation of the Consequent.

If you repent, then only are you forgiven.

You are forgiven.

∴ You repent.

CHAPTER XXI.

Of the Reduction of the Partly Conjunctive Syllogism.

§ 744. Such syllogisms as those just treated of, if syllogisms they are to be called, have a major and a middle term visible to the eye, but appear to be destitute of a minor. The missing minor term is however supposed to be latent in the transition from the conjunctive to the simple form of proposition. When we say 'A is B,' we are taken to mean, 'As a matter of fact, A is B' or 'The actual state of the case is that A is B.' The insertion therefore of some such expression as 'The case in hand,' or 'This case,' is, on this view, all that is wanted to complete the form of the syllogism. When reduced in this manner to the simple type of argument, it will be found that the constructive conjunctive conforms to the first figure and the destructive conjunctive to the second.

Constructive Mood. Barbara.

If A is B, C is D. \ / All cases of A being B are cases of

   \ = / C being D.

A is B. / \ This is a case of A being B.

∴ C is D. / \ ∴ This is a case of C being D.

Destructive Mood. Camestres.

If A is B, C is D. \ / All cases of A being B are cases of

   \ = / C being D.

C is not D. / \ This is not a case of C being D.

∴ A is not B. / \ ∴ This is not a case of A being B.

§ 745. It is apparent from the position of the middle term that the constructive conjunctive must fall into the first figure and the destructive conjunctive into the second. There is no reason, however, why they should be confined to the two moods, Barbara and Carnestres. If the inference is universal, whether as general or singular, the mood is Barbara or Carnestres; if it is particular, the mood is Darii or Baroko.

   Barbara. Camestres.

If A is B, C is always D. \ If A is B, C is always D. \

A is always B. \ C is never D. \
∴ C is always D. \ ∴ A is never B. \
　　　　| |
If A is B, C is always D. / If A is B, C is always D. /
A is in this case B. / C is not in this case D. /
∴ C is in this case D. / ∴ A is not in this case B. /
　Darii. Baroko.
　If A is B, C is always D. If A is B, C is never D.
A is sometimes B. C is sometimes not D.
∴ C is sometimes D. ∴ A is sometimes not B.

§ 746. The remaining moods of the first and second figure are obtained by taking a negative proposition as the consequent in the major premiss.
　　Celarent. Ferio.
If A is B, C is never D. If A is B, C is never D.
A is always B. A is sometimes B.
∴ C is never D. ∴ C is sometimes not D.
　Cesare. Festino.
If A is B, C is never D. If A is B, C is never D.
C is always D. C is sometimes D.
∴ A is never B. ∴ A is sometimes not B.

§ 747. As the partly conjunctive syllogism is thus reducible to the simple form, it follows that violations of its laws must correspond with violations of the laws of simple syllogism. By our throwing the illicit moods into the simple form it will become apparent what fallacies are involved in them.
　Denial of Anteceded.
　If A is B, C is D. \ / All cases of A being B are cases of C
　　　　\ = / being D.
A is not B. / \ This is not a case of A being B.
∴ C is not D. / \ ∴ This is not a case of C being D.
　Here we see that the denial of the antecedent amounts to illicit process of the major term.

§ 7481 *Affirmation of Consequent.*
　If A is B, C is D. \ / All Cases of A being B are cases of C
　　　　| = | being D.
C is D. / \ This is a case of C being D.
　Here we see that the affirmation of the consequent amounts to undistributed middle.

§ 749. If we confine ourselves to the special rules of the four figures, we see that denial of the antecedent involves a negative minor in the first figure, and affirmation of the consequent two affirmative premisses in the second. Or, if the consequent in the major premiss were itself negative, the affirmation of it would amount to the fallacy of two negative premisses. Thus—
　If A is B, C is not D. \ / No cases of A being B are cases of C
　　　　| = | being D.
C is not D. / \ This is not a case of C being D.

§ 750. The positive side of the canon of the conjunctive syllogism—'To affirm the antecedent is to affirm the consequent,' corresponds with the Dictum de Omni. For whereas something (viz. C being D) is affirmed in the major of all conceivable cases of A being B, the same is affirmed in the conclusion of something which is included therein, namely, 'this case,' or 'some cases,' or even 'all actual cases.'

§ 751. The negative side—'to deny the consequent is to deny the antecedent'—corresponds with the Dictum de Diverse (§ 643). For whereas in the major all conceivable cases of A being B are included in C being D, in the minor 'this case,' or 'some cases,' or even 'all actual cases' of C being D, are excluded from the same notion.

§ 752. The special characteristic of the partly conjunctive syllogism lies in the transition from hypothesis to fact. We might lay down as the appropriate axiom of this form of argument, that 'What is true in the abstract is true—in the concrete,' or 'What is true in theory is also true in fact,' a proposition which is apt to be neglected or denied. But this does not vitally distinguish it from the ordinary syllogism. For though in the latter we think rather of the transition from a general truth to a particular application of it, yet at bottom a general truth is nothing but a hypothesis resting upon a slender basis of observed fact. The proposition 'A is B' may be expressed in the form 'If A is, B is.' To say that 'All men are mortal' may be interpreted to mean that 'If we find in any subject the attributes of humanity, the attributes of mortality are sure to accompany them.'

CHAPTER XXII.

Of the Partly Conjunctive Syllogism regarded as an Immediate Inference.

§ 753. It is the assertion of fact in the minor premiss, where we have the application of an abstract principle to a concrete instance, which alone entitles the partly conjunctive syllogism to be regarded as a syllogism at all. Apart from this the forms of semi-conjunctive reasoning run at once into the moulds of immediate inference.

§ 754. The constructive mood will then be read in this way—

If A is B, C is D,

∴ A being B, C is D.

reducing itself to an instance of immediate inference by subaltern opposition—

Every case of A being B, is a case of C being D.

∴ Some particular case of A being B is a case of C being D.

§ 755. Again, the destructive conjunctive will read as follows—

If A is B, C is D,

∴ C not being D, A is not B.

which is equivalent to

All cases of A being B are cases of C being D.

∴ Whatever is not a case of C being D is not a case of A being B.

∴ Some particular case of C not being D is not a case of A being B.

But what is this but an immediate inference by contraposition, coming under the formula

All A is B,

∴ All not-B is not-A,

and followed by Subalternation?

§ 756. The fallacy of affirming the consequent becomes by this mode of treatment an instance of the vice of immediate inference known as the simple conversion of an A proposition. 'If A is B, C is D' is not convertible with 'If C is D, A is B' any more than 'All A is B' is convertible with 'All B is A.'

§ 757. We may however argue in this way

If A is B, C is D,

C is D,

∴ A may be B,

which is equivalent to saying,
    When A is B, C is always D,
∴ When C is D, A is sometimes B,
    and falls under the legitimate form of conversion of A per accidens—
    All cases of A being B are cases of C being D.
∴ Some cases of C being D are cases of A being B.
    § 758. The fallacy of denying the antecedent assumes the following form—
    If A is B, C is D,
∴ If A is not B, C is not D,
    equivalent to—
    All cases of A being B are cases of C being D.
∴ Whatever is not a case of A being B is not a case of C being D.
    This is the same as to argue—
    All A is B,
∴ All not-A is not-B,
an erroneous form of immediate inference for which there is no special name, but which involves the vice of simple conversion of A, since 'All not-A is not-B' is the contrapositive, not of 'All A is B,' but of its simple converse 'All B is A.'
    § 759. The above-mentioned form of immediate inference, however (namely, the employment of contraposition without conversion), is valid in the case of the U proposition; and so also is simple conversion. Accordingly we are able, as we have seen, in dealing with a proposition of that form, both to deny the antecedent and to assert the consequent with impunity—
    If A is B, then only C is D,
∴ A not being B, C is not D;
    and again, C being D, A must be B.

CHAPTER XXIII.

Of the Disjunctive Syllogism.
    § 760. Roughly speaking, a Disjunctive Syllogism results from the combination of a disjunctive with a simple premiss. As in the preceding form, the complex proposition is regarded as the major premiss, since it lays down a hypothesis, which is applied to fact in the minor.
    § 761. The Disjunctive Syllogism may be exactly defined as follows—
    A complex syllogism, which has for its major premiss a disjunctive proposition, either the antecedent or consequent of which is in the minor premiss simply affirmed or denied.
    § 762. Thus there are four types of disjunctive syllogism possible.
Constructive Moods.
    (1) Either A is B or C is D. (2) Either A is B or C is D.
    A is not B. C is not D.
∴ C is D. ∴ A is B.
    Either death is annihilation or we are immortal.
Death is not annihilation.
∴ We are immortal.

Either the water is shallow or the boys will be drowned.
The boys are not drowned.
∴. The water is shallow.

Destructive Moods.

(3) Either A is B or C is D. (4) Either A is B or C is D.
A is B. C is D.
∴. C is not D. ∴. A is not B.

§ 763. Of these four, however, it is only the constructive moods that are formally conclusive. The validity of the two destructive moods is contingent upon the kind of alternatives selected. If these are such as necessarily to exclude one another, the conclusion will hold, but not otherwise. They are of course mutually exclusive whenever they embody the result of a correct logical division, as 'Triangles are either equilateral, isosceles or scalene.' Here, if we affirm one of the members, we are justified in denying the rest. When the major thus contains the dividing members of a genus, it may more fitly be symbolized under the formula, 'A is either B or C.' But as this admits of being read in the shape, 'Either A is B or A is C,' we retain the wider expression which includes it. Any knowledge, however, which we may have of the fact that the alternatives selected in the major are incompatible must come to us from material sources; unless indeed we have confined ourselves to a pair of contradictory terms (A is either B or not-B). There can be nothing in the form of the expression to indicate the incompatibility of the alternatives, since the same form is employed when the alternatives are palpably compatible. When, for instance, we say, 'A successful student must be either talented or industrious,' we do not at all mean to assert the positive incompatibility of talent and industry in a successful student, but only the incompatibility of their negatives—in other words, that, if both are absent, no student can be successful. Similarly, when it is said, 'Either your play is bad or your luck is abominable,' there is nothing in the form of the expression to preclude our conceiving that both may be the case.

§ 764. There is no limit to the number of members in the disjunctive major. But if there are only two alternatives, the conclusion will be a simple proposition; if there are more than two, the conclusion will itself be a disjunctive. Thus—

Either A is B or C is D or E is F or G is H.
E is not F.
∴. Either A is B or C is D or G is H.

§ 765. The Canon of the Disjunctive Syllogism may be laid down as follows—
To deny one member is to affirm the rest, either simply or disjunctively; but from affirming any member nothing follows.

CHAPTER XXIV.

Of the Reduction of the Disjunctive Syllogism.

§ 766. We have seen that in the disjunctive syllogism the two constructive moods alone are formally valid. The first of these, namely, the denial of the antecedent, will in all cases give a simple syllogism in the first figure; the second of them, namely, the denial of the consequent, will in all cases give a simple syllogism in the second figure.

Denial of Antecedent = Barbara.

Either A is B or C is D.

A is not B.

.'.C is D

is equal to

If A is not B, C is D.

A is not B.

.'. C is D.

is equal to

All cases of A not being B are cases of C being D.

This is a case of A not being B.

.'. This is a case of C being D.

Denial of Consequent = Camestres.

Either A is E or C is D.

C is not D.

.'. A is B.

is equal to

If A is not B, C is D.

C is not D.

.'. A is B.

is equal to

All cases of A not being B are cases of C being D.

This is not a case of C being D.

.'. This is not a case of A being B.

§ 767. The other moods of the first and second figures can be obtained by varying the quality of the antecedent and consequent in the major premiss and reducing the quantity of the minor.

§ 768. The invalid destructive moods correspond with the two invalid types of the partly conjunctive syllogism, and have the same fallacies of simple syllogism underlying them. Affirmation of the antecedent of a disjunctive is equivalent to the semi-conjunctive fallacy of denying the antecedent, and therefore involves the ordinary syllogistic fallacy of illicit process of the major.

Affirmation of the consequent of a disjunctive is equivalent to the same fallacy in the semi-conjunctive form, and therefore involves the ordinary syllogistic fallacy of undistributed middle.

Affirmation of Antecedent = *Illicit Major.*

Either A is B or C is D.

A is B.

.'. C is not D.

is equal to

If A is not B, C is D.

A is B.

.'. C is not D.

is equal to

All cases of A not being B are cases of C being D.

This is not a case of A not being B.

.'. This is not a case of C not being D.

Affirmation of Consequent = *Undistributed Middle.*

Either A is B or C is D.

C is D.

is equal to

If A is not B, C is D.

C is D.

is equal to

    All cases of A not being B are cases of C being D.

    This is a case of C being D.

§ 769. So far as regards the consequent, the two species of complex reasoning hitherto discussed are identical both in appearance and reality. The apparent difference of procedure in the case of the antecedent, namely, that it is affirmed in the partly conjunctive, but denied in the disjunctive syllogism, is due merely to the fact that in the disjunctive proposition the truth of the consequent is involved in the falsity of the antecedent, so that the antecedent being necessarily negative, to deny it in appearance is in reality to assert it.

CHAPTER XXV.

The Disjunctive Syllogism regarded as an Immediate Inference.

§ 770. If no stress be laid on the transition from disjunctive hypothesis to fact, the disjunctive syllogism will run with the same facility as its predecessor into the moulds of immediate inference.

§ 771.

Denial of Antecedent. Subalternation.

    Either A is B or C is D, Every case of A not being B

        is a case of C being D.

∴ A not being B, C is D. ∴ Some case of A not being B

        is a case of C being D.

§ 772.

Denial of Consequent. Conversion by Contraposition

        + Subalternation.

    Either A is B or C is D. All cases of A not being B

        are cases of C being D.

∴ C not being D, A is B ∴ All cases of C not being D are

        cases of A being B.

        ∴ Some case of C not being D is

        a case of A being B.

§ 773. Similarly the two invalid types of disjunctive syllogism will be found to coincide with fallacies of immediate inference.

§ 774.

Affirmation of Antecedent. Contraposition without

        Conversion.

    Either A is B or C is D. All cases of A not being B are

        cases of C being D.

∴ A being B, C is not D ∴ All cases of A being B are

        cases of C not being D.

§ 775. The affirmation of the antecedent thus comes under the formula—

    All not-A is B,

∴ All A is not-B,

a form of inference which cannot hold except where A and B are known to be incompatible. Who, for instance, would assent to this?—

All non-boating men play cricket.

∴ All boating men are non-cricketers.

§ 776.

Affirmation of Consequent. Simple Conversion of A.

Either A is B or C is D. All cases of A not being B are
cases of C being D.

∴C being D, A is not B. ∴ All cases of C being D are
cases of A not being B.

§ 777. We may however argue in this way—

Conversion of A per accidens.

Either A is B or C is D. All cases of A not being B
are cases of C being D.

∴ C being D, A is sometimes B. ∴ Some cases of C being D are
cases of A not being B.

The men who pass this examination must have either talent or industry.

∴ Granting that they are industrious, they may be without talent.

CHAPTER XXVI.

Of the Mixed Form of Complex Syllogism.

§ 778. Under this head are included all syllogisms in which a conjunctive is combined with a disjunctive premiss. The best known form is

The Dilemma.

§ 779. The Dilemma may be defined as—

A complex syllogism, having for its major premiss a conjunctive proposition
with more than one antecedent, or more than one consequent, or both, which
(antecedent or consequent) the minor premiss disjunctively affirms or denies.

§ 780. It will facilitate the comprehension of the dilemma, if the following three points are borne in mind—

(1) that the dilemma conforms to the canon of the partly conjunctive syllogism,
and therefore a valid conclusion can be obtained only by affirming the antecedent
or denying the consequent;

(2) that the minor premiss must be disjunctive;

(3) that if only the antecedent be more than one, the conclusion will be a simple
proposition; but if both antecedent and consequent be more than one, the
conclusion will itself be disjunctive.

§ 781. The dilemma, it will be seen, differs from the partly conjunctive syllogism chiefly in the fact of having a disjunctive affirmation of the antecedent or denial of the consequent in the minor, instead of a simple one. It is this which constitutes the essence of the dilemma, and which determines its possible varieties. For if only the antecedent or only the consequent be more than one, we must, in order to obtain a disjunctive minor, affirm the antecedent or deny the consequent respectively; whereas, if there be more than one of both, it is open to us to take either course. This gives us four types of dilemma.

§ 782.

(1). *Simple Constructive.*

If A is B or C is D, E is F.
Either A is B or C is D.
∴. E is F.
(2). *Simple Destructive.*
  If A is B, C is D and E is F.
Either C is not D or E is not F.
∴. A is not B.
(3). *Complex Constructive.*
  If A is B, C is D; and if E is F, G is H.
Either A is B or E is F.
∴. Either C is D or G is H.
(4). *Complex Destructive.*
  If A is B, C is D; and if E is F, G is H.
Either C is not D or G is not H.
∴. Either A is not B or E is not F.
§ 783.
(1). *Simple Constructive.*
  If she sinks or if she swims, there will be an end of her.
She must either sink or swim.
∴. There will be an end of her.
(2). *Simple Destructive.*
  If I go to Town, I must pay for my ticket and pay my hotel bill.
Either I cannot pay for my ticket or I cannot pay my hotel bill.
∴. I cannot go to Town.
(3). *Complex Constructive.*
  If I stay in this room, I shall be burnt to death, and if I jump
out of the window, I shall break my neck.
I must either stay in the room or jump out of the window.
∴. I must either be burnt to death or break my neck.
(4). *Complex Destructive.*
  If he were clever, he would see his mistake; and
if he were candid, he would acknowledge it.
Either he does not see his mistake or he will not acknowledge it.
∴. Either he is not clever or he is not candid.

§ 784. It must be noticed that the simple destructive dilemma would not admit of a disjunctive consequent. If we said,

If A is B, either C is D or E is F,

Either C is not D or E is not F,

we should not be denying the consequent. For 'E is not F' would make it true that C is D, and 'C is not D' would make it true that E is F; so that in either case we should have one of the alternatives true, which is just what the disjunctive form 'Either C is D or E is F' insists upon.

§ 785. In the case of the complex constructive dilemma the several members, instead of being distributively assigned to one another, may be connected together as a whole—thus—

If either A is B or E is F, either C is D or G is H.

Either A is B or E is F.

∴. Either C is D or G is H.

In this shape the likeness of the dilemma to the partly conjunctive syllogism is more immediately recognisable. The major premiss in this shape is vaguer than in the former. For each antecedent has now a disjunctive choice of consequents, instead of being limited to one. This vagueness, however, does not affect the conclusion. For, so long as the conclusion is established, it does not matter from which members of the major its own members flow.

§ 786. It must be carefully noticed that we cannot treat the complex destructive dilemma in the same way.

If either A is B or E is F, either C is D or G is H.
Either C is not D or G is not H.

Since the consequents are no longer connected individually with the antecedents, a disjunctive denial of them leaves it still possible for the antecedent as a whole to be true. For 'C is not D' makes it true that G is H, and 'G is not H' makes it true that C is D. In either case then one is true, which is all that was demanded by the consequent of the major. Hence the consequent has not really been denied.

§ 787. For the sake of simplicity we have limited the examples to the case of two antecedents or consequents. But we may have as many of either as we please, so as to have a Trilemma, a Tetralemma, and so on.

TRILEMMA.

If A is B, C is D; and if E is F, G is H; and if K is L, M is N.
Either A is B or E is F or K is L.
∴ Either C is D or G is H or K is L.

§ 788. Having seen what the true dilemma is, we shall now examine some forms of reasoning which resemble dilemmas without being so.

§ 789. This, for instance, is not a dilemma—

If A is B or if E is F, C is D.
But A is B and E is F.
∴ C is D.

If he observes the sabbath or if he refuses to eat pork, he is a
Jew.
But he both observes the sabbath and refuses to eat pork.
∴ He is a Jew.

What we have here is a combination of two partly conjunctive syllogisms with the same conclusion, which would have been established by either of them singly. The proof is redundant.

§ 790. Neither is the following a dilemma—

If A is B, C is D and E is F.
Neither C is D nor E is F.
∴ A is not B.

If this triangle is equilateral, its sides and its angles will be
equal.
But neither its sides nor its angles are equal.
∴ It is not equilateral.

This is another combination of two conjunctive syllogisms, both pointing to the same conclusion. The proof is again redundant. In this case we have the consequent denied in both, whereas in the former we had the antecedent affirmed. It is only for convenience that such arguments as these are thrown into the form of a single syllogism. Their real distinctness may be seen from the fact that we here deny each proposition separately, thus making two independent statements—C is not D and E is not F. But in the true instance of the simple destructive dilemma, what we deny is not the truth of the two propositions contained in the consequent, but their compatibility; in other words we make a disjunctive denial.

§ 791. Nor yet is the following a dilemma—

If A is B, either C is D or E is F.
Neither C is D nor E is F.
∴ A is not B.

If the barometer falls there will be either wind or rain.

There is neither wind nor rain.

.'. The barometer has not fallen.

What we have here is simply a conjunctive major with the consequent denied in the minor. In the consequent of the major it is asserted that the two propositions, 'C is D' and 'E is F' cannot both be false; and in the minor this is denied by the assertion that they are both false.

§ 792. A dilemma is said to be rebutted or retorted, when another dilemma is made out proving an opposite conclusion. If the dilemma be a sound one, and its premisses true, this is of course impossible, and any appearance of contradiction that may present itself on first sight must vanish on inspection. The most usual mode of rebutting a dilemma is by transposing and denying the consequents in the major—

If A is B, C is D; and if E is F, G is H.

Either A is B or E is F.

.'. Either C is D or G is H.

The same rebutted—

If A is B, G is not H; and if E is F, C is not D.

Either A is B or E is F.

.'. Either G is not H or C is not D.

= Either C is not D or G is not H.

§ 793. Under this form comes the dilemma addressed by the Athenian mother to her son—'Do not enter public life: for, if you say what is just, men will hate you; and, if you say what is unjust, the gods will hate you' to which the following retort was made—'I ought to enter public life: for, if I say what is just, the gods will love me; and, if I say what is unjust, men will love me.' But the two conclusions here are quite compatible. A man must, on the given premisses, be both hated and loved, whatever course he takes. So far indeed are two propositions of the form

Either C is D or G is H,

and Either C is not D or G is not H,

from being incompatible, that they express precisely the same thing when contradictory alternatives have been selected, e.g.—

Either a triangle is equilateral or non-equilateral.

Either a triangle is non-equilateral or equilateral.

§ 794. Equally illusory is the famous instance of rebutting a dilemma contained in the story of Protagoras and Euathlus (Aul. Gell. Noct. Alt. v. 10), Euathlus was a pupil of Protagoras in rhetoric. He paid half the fee demanded by his preceptor before receiving lessons, and agreed to pay the remainder when he won his first case. But as he never proceeded to practise at the bar, it became evident that he meant to bilk his tutor. Accordingly Protagoras himself instituted a law-suit against him, and in the preliminary proceedings before the jurors propounded to him the following dilemma—'Most foolish young man, whatever be the issue of this suit, you must pay me what I claim: for, if the verdict be given in your favour, you are bound by our bargain; and if it be given against you, you are bound by the decision of the jurors.' The pupil, however, was equal to the occasion, and rebutted the dilemma as follows. 'Most sapient master, whatever be the issue of this suit, I shall not pay you what you claim: for, if the verdict be given in my favour, I am absolved by the decision of the jurors; and, if it be given against me, I am absolved by our bargain.' The jurors are said to have been so puzzled by the conflicting plausibility of the arguments that they adjourned the case till the Greek Kalends. It is evident, however, that a grave injustice was thus done to Protagoras. His dilemma was really invincible. In the counter-dilemma of Euathlus we are meant to infer that Protagoras would actually lose his fee, instead of merely getting it in one way rather than another. In either case he would both get and lose his fee, in the sense of getting it on one plea, and not getting it on another: but in neither case would he actually lose it.

§ 795. If a dilemma is correct in form, the conclusion of course rigorously follows: but a material fallacy often underlies this form of argument in the tacit assumption that the alternatives offered in the minor constitute an exhaustive division. Thus the dilemma 'If pain is severe, it will be brief; and if it last long it will be slight,' &c., leaves out of sight the unfortunate fact that pain may both be severe and of long continuance. Again the following dilemma—

If students are idle, examinations are unavailing; and, if
they are industrious, examinations are superfluous,
Students are either idle or industrious,
.'. Examinations are either unavailing or superfluous,

is valid enough, so far as the form is concerned. But the person who used it would doubtless mean to imply that students could be exhaustively divided into the idle and the industrious. No deductive conclusion can go further than its premises; so that all that the above conclusion can in strictness be taken to mean is that examinations are unavailing, when students are idle, and superfluous, when they are industrious—which is simply a reassertion as a matter of fact of what was previously given as a pure hypothesis.

CHAPTER XXVII.

Of the Reduction of the Dilemma.

§ 796. As the dilemma is only a peculiar variety of the partly conjunctive syllogism, we should naturally expect to find it reducible in the same way to the form of a simple syllogism. And such is in fact the case. The constructive dilemma conforms to the first figure and the destructive to the second.

1) *Simple Constructive Dilemma.*

Barbara.

If A is B or if E is F, C is D. All cases of either A being B or E
        being F are cases of C being D.
Either A is B or E is F. All actual cases are cases of either
        A being B OP E being F.
.'. C is D. .'. All actual cases are cases of C
        being D.

(2) *Simple Destructive.*

Camstres.

If A is B, C is D and E is F. All cases of A being B are cases of
        C being D and E being F.
Either C is not D or E is not F. No actual cases are cases of C being
        D and E being F.
.'. A is not B. .'. No actual cases are cases of A
        being B.

(3) *Complex Constructive.*

Barbara.

If A is B, C is D; and if E is F, All cases of either A being B or
G is H. being F are cases of either C being
        D or G being H.

Either A is B or E is F. All actual cases are cases of either A
                being B or E being F.
.'. Either C is D or G is H. .'. All actual cases are cases of either C
                being D or G being H.

(4) *Complex Destructive.*

If A is B, C is D; and if E is F, All cases of A being B and E being F
G is H. are cases of C being D and G
                being H.
Either C is not D Or G is No actual cases are cases of C being
  not H D and G being H.
Either A is not B or E is No actual cases are cases of A being
not F. B and E being F.

§ 797. There is nothing to prevent our having Darii, instead of Barbara, in the constructive form, and Baroko, instead of Camestres, in the destructive. As in the case of the partly conjunctive syllogism the remaining moods of the first and second figure are obtained by taking a negative proposition as the consequent of the major premiss, e.g.—

Simple Constructive. Celarent or Ferio.

If A is B or if E is F, C is not D No cases of either A being B or E
                being F are cases of C being D.
Either A is B or E is F. All (or some) actual cases are cases of
                either A being B or E being F
.'. C is not D. .'. All (or some) actual cases are not
                cases of C being D.

CHAPTER XXVIII.

Of the Dilemma regarded as an Immediate Inference.

§ 798. Like the partly conjunctive syllogism, the dilemma can be expressed under the forms of immediate inference. As before, the conclusion in the constructive type resolves itself into the subalternate of the major itself, and in the destructive type into the subalternate of its contrapositive. The simple constructive dilemma, for instance, may be read as follows—

If either A is B or E is F, C is D,
.'. Either A being B or E being F, C is D,

which is equivalent to

Every case of either A being B or E being F is a case of C being D.
.'. Some case of either A being B or E being F is a case of C being D.

The descent here from 'every' to 'some' takes the place of the transition from hypothesis to fact.

§ 799. Again the complex destructive may be read thus—

If A is B, C is D; and if E is F, G is H,
.'. It not being true that C is D and G is H, it is not
  true that A is B and E is F,

which may be resolved into two steps of immediate inference, namely, conversion by contraposition followed by subalternation—

All cases of A being B and E being F are cases of C being D and G being H.

∴. Whatever is not a case of C being D and G being H is not a case of A being B and E being F.

∴. Some case which is not one of C being D and G being H is not a case of A being B and E being F.

CHAPTER XXIX.

Of Trains of Reasoning.

§ 800. The formal logician is only concerned to examine whether the conclusion duly follows from the premisses: he need not concern himself with the truth or falsity of his data. But the premisses of one syllogism may themselves be conclusions deduced from other syllogisms, the premisses of which may in their turn have been established by yet earlier syllogisms. When syllogisms are thus linked together we have what is called a Train of Reasoning.

§ 801. It is plain that all truths cannot be established by reasoning. For the attempt to do so would involve us in an infinite regress, wherein the number of syllogisms required would increase at each step in a geometrical ratio. To establish the premisses of a given syllogism we should require two preceding syllogisms; to establish their premisses, four; at the next step backwards, eight; at the next, sixteen; and so on ad infinitum. Thus the very possibility of reasoning implies truths that are known to us prior to all reasoning; and, however long a train of reasoning may be, we must ultimately come to truths which are either self-evident or are taken for granted.

§ 802. Any syllogism which establishes one of the premisses of another is called in reference to that other a Pro-syllogism, while a syllogism which has for one of its premisses the conclusion of another syllogism is called in reference to that other an Epi-syllogism.

The Epicheirema.

§ 803. The name Epicheirema is given to a syllogism with one or both of its premisses supported by a reason. Thus the following is a double epicheirema—

All B is A, for it is E.
All C is B, for it is F.
∴. All C is A.

All virtue is praiseworthy, for it promotes the general welfare.
Generosity is a virtue, for it prompts men to postpone self to others.
∴. Generosity is praiseworthy.

§ 804. An epicheirema is said to be of the first or second order according as the major or minor premiss is thus supported. The double epicheirema is a combination of the two orders.

§ 805. An epicheirema, it will be seen, consists of one syllogism fully expressed together with one, or, it may be, two enthymemes (§ 557). In the above instance, if the reasoning which supports the premisses were set forth at full length, we should have, in place of the enthymemes, the two following pro-syllogisms—

(i) All E is A.
All B is E.
∴. All B is A.

Whatever promotes the general welfare is praiseworthy.
Every virtue promotes the general welfare.
.'. Every virtue is praiseworthy.
   (2) All F is B.
   All C is F.
   .'. All C is B.
Whatever prompts men to postpone self to others is a virtue.
Generosity prompts men to postpone self to others.
.'. Generosity is a virtue.

§ 806. The enthymemes in the instance above given are both of the first order, having the major premiss suppressed. But there is nothing to prevent one or both of them from being of the second order—

All B is A, because all F is.
All C is B, because all F is.
.'. All C is A.

All Mahometans are fanatics, because all Monotheists are.
These men are Mahometans, because all Persians are.
.'. These men are fanatics.

Here it is the minor premiss in each syllogism that is suppressed, namely,

(1) All Mahometans are Monotheists.
(2) These men are Persians.

The Sorites.

§ 807. The Sorites is the neatest and most compendious form that can be assumed by a train of reasoning.

§ 808. It is sometimes more appropriately called the chain-argument, and map be defined as—

A train of reasoning, in which one premiss of each epi-syllogism is
supported by a pro-syllogism, the other being taken for granted.

This is its inner essence.

§ 809. In its outward form it may be described as—A series of propositions, each of which has one term in common with that which preceded it, while in the conclusion one of the terms in the last proposition becomes either subject or predicate to one of the terms in the first.

§ 810. A sorites may be either—
(1) Progressive,
or (2) Regressive.

Progressive Sorites.

   All A is B.
All B is C.
All C is D.
All D is E.
.'. All A is E.

Regressive Sorites.

   All D is E.
All C is D.
All B is C.
All A is B.
.'. All A is E.

§ 811. The usual form is the progressive; so that the sorites is commonly described as a series of propositions in which the predicate of each becomes the subject of the next, while in the conclusion the last predicate is affirmed or denied of the first subject. The regressive form, however, exactly reverses these attributes; and would require to be described as a series of propositions, in which the subject of each becomes the predicate of the next, while in the conclusion the first predicate is affirmed or denied of the last subject.

§ 812. The regressive sorites, it will be observed, consists of the same propositions as the progressive one, only written in reverse order. Why then, it may be asked, do we give a special name to it, though we do not consider a syllogism different, if the minor premiss happens to precede the major? It is because the sorites is not a mere series of propositions, but a compressed train of reasoning; and the two trains of reasoning may be resolved into their component syllogisms in such a manner as to exhibit a real difference between them.

§ 813. The Progressive Sorites is a train of reasoning in which the minor premiss of each epi-syllogism is supported by a pro-syllogism, while the major is taken for granted.

§ 814. The Regressive Sorites is a train of reasoning in which the major premiss of each epi-syllogism is supported by a pro-syllogism, while the minor is taken for granted.

Progressive Sorites.
(i) All B is C.
All A is B.
∴ All A is C.
(2) All C is D.
All A is C.
∴ All A is D.
(3) All D is E.
All A is D.
∴ All A is E.

Regressive Sorites.
(1) All D is E.
All C is D.
∴ All C is E.
(2) All C is E.
All B is C.
∴ All B is E.
(3) All B is E.
All A is B.
∴ All A is E.

§ 815. Here is a concrete example of the two kinds of sorites, resolved each into its component syllogisms—

Progressive Sorites.
All Bideford men are Devonshire men.
All Devonshire men are Englishmen.
All Englishmen are Teutons.
All Teutons are Aryans.
∴ All Bideford men are Aryans.
(1) All Devonshire men are Englishmen.
All Bideford men are Devonshire men.
∴ All Bideford men are Englishmen.
(2) All Englishmen are Teutons.
All Bideford men are Englishmen.
∴ All Bideford men are Teutons.
(3) All Teutons are Aryans.
All Bideford men are Teutons.
∴ All Bideford men are Aryans.

Regressive Sorites.

All Teutons are Aryans.
All Englishmen are Teutons.
All Devonshiremen are Englishmen.
All Bideford men are Devonshiremen.
.'. All Bideford men are Aryans.
   (1) All Teutons are Aryans.
   All Englishmen are Teutons.
   .'. All Englishmen are Aryans.
   (2) All Englishmen are Aryans.
   All Devonshiremen are Englishmen.
   .'. All Devonshiremen are Aryans.
   (3) All Devonshiremen are Aryans.
   All Bideford men are Devonshiremen.
   .'. All Bideford men are Aryans.

§ 816. When expanded, the sorites is found to contain as many syllogisms as there are propositions intermediate between the first and the last. This is evident also on inspection by counting the number of middle terms.

§ 817. In expanding the progressive form we have to commence with the second proposition of the sorites as the major premiss of the first syllogism. In the progressive form the subject of the conclusion is the same in all the syllogisms; in the regressive form the predicate is the same. In both the same series of means, or middle terms, is employed, the difference lying in the extremes that are compared with one another through them.

[Illustration]

§ 818. It is apparent from the figure that in the progressive form we work from within outwards, in the regressive form from without inwards. In the former we first employ the term 'Devonshiremen' as a mean to connect 'Bideford men' with 'Englishmen'; next we employ 'Englishmen' as a mean to connect the same subject 'Bideford men' with the wider term 'Teutons'; and, lastly, we employ 'Teutons' as a mean to connect the original subject 'Bideford men' with the ultimate predicate 'Ayrans.'

§ 819. Reversely, in the regressive form we first use 'Teutons' as a mean whereby to bring 'Englishmen' under 'Aryans'; next we use 'Englishmen' as a mean whereby to bring 'Devonshiremen' under the dame predicate 'Aryans'; and, lastly, we use 'Devonshiremen' as a mean whereby to bring the ultimate subject 'Bideford men' under the original predicate 'Aryans.'

§ 820. A sorites may be either Regular or Irregular.

§ 821. In the regular form the terms which connect each proposition in the series with its predecessor, that is to say, the middle terms, maintain a fixed relative position; so that, if the middle term be subject in one, it will always be predicate in the other, and vice versâ. In the irregular form this symmetrical arrangement is violated.

§ 822. The syllogisms which compose a regular sorites, whether progressive or regressive, will always be in the first figure.

In the irregular sorites the syllogisms may fall into different figures.

§ 823. For the regular sorites the following rules may be laid down.

   (1) Only one premiss can be particular, namely, the first, if the sorites be progressive, the last, if it be regressive.

   (2) Only one premiss can be negative, namely, the last, if the sorites be progressive, the first, if it be regressive.

§ 824. *Proof of the Rules for the Regular Sorites.*

   (1) In the progressive sorites the proposition which stands first is the only one which appears as a minor premiss in the expanded form. Each of the others is used in its turn as a major. If any proposition, therefore, but the first were particular,

there would be a particular major, which involves undistributed middle, if the minor be affirmative, as it must be in the first figure.

In the regressive sorites, if any proposition except the last were particular, we should have a particular conclusion in the syllogism in which it occurred as a premiss, and so a particular major in the next syllogism, which again is inadmissible, as involving undistributed middle.

(2) In the progressive sorites, if any premiss before the last were negative, we should have a negative conclusion in the syllogism in which it occurs. This would necessitate a negative minor in the next syllogism, which is inadmissible in the first figure, as involving illicit process of the major.

In the regressive sorites the proposition which stands first is the only one which appears as a major premiss in the expanded form. Each of the others is used in its turn as a minor. If any premiss, therefore, but the first were negative, we should have a negative minor in the first figure, which involves illicit process of the major.

§ 825. The rules above given do not apply to the irregular sorites, except so far as that only one premiss can be particular and only one negative, which follows from the general rules of syllogism. But there is nothing to prevent any one premiss from being particular or any one premiss from being negative, as the subjoined examples will show. Both the instances chosen belong to the progressive order of sorites.

(1) Barbara.
All B is A.
All C is B.
All C is A.
   All B is A.
All C is B.
Some C is D.
All D is E
   .'. Some A is E
[Illustration]
(2) Disamis.
Some C is D.
All C is A.
Some A is D.
(3) Darii.
All D is E
Some A is D.
Some A is E.
(1) Barbara.
All B is C.
All A is B.
All A is C.
   All A is B.
All B is C.
No D is C.
All E is D.
   .'. No A is E.
[Illustration]
(2) Cesare.
No D is C.
All A is C.
   .'. No A is D.

(3) Camestres.
All E is D.
No A is D.
∴ No A is E.

§ 826. A chain argument may be composed consisting of conjunctive instead of simple propositions. This is subject to the same laws as the simple sorites, to which it is immediately reducible.

Progressive. Regressive.
If A is B, C is D. If E is F, G is H.
If C is D, E is F. If C is D, E is F.
If E is F, G is H. If A is B, C is D.
∴ If A is B, G is H. ∴ If A is B, G is H.

CHAPTER XXX.

Of Fallacies.

§ 827. After examining the conditions on which correct thoughts depend, it is expedient to classify some of the most familiar forms of error. It is by the treatment of the Fallacies that logic chiefly vindicates its claim to be considered a practical rather than a speculative science. To explain and give a name to fallacies is like setting up so many sign-posts on the various turns which it is possible to take off the road of truth.

§ 828. By a fallacy is meant a piece of reasoning which appears to establish a conclusion without really doing so. The term applies both to the legitimate deduction of a conclusion from false premises and to the illegitimate deduction of a conclusion from any premises. There are errors incidental to conception and judgement, which might well be brought under the name; but the fallacies with which we shall concern ourselves are confined to errors connected with inference.

§ 829. When any inference leads to a false conclusion, the error may have arisen either in the thought itself or in the signs by which the thought is conveyed. The main sources of fallacy then are confined to two—

(1) thought,
(2) language.

§ 830. This is the basis of Aristotle's division of fallacies, which has not yet been superseded. Fallacies, according to him, are either in the language or outside of it. Outside of language there is no source of error but thought. For things themselves do not deceive us, but error arises owing to a misinterpretation of things by the mind. Thought, however, may err either in its form or in its matter. The former is the case where there is some violation of the laws of thought; the latter whenever thought disagrees with its object. Hence we arrive at the important distinction between Formal and Material fallacies, both of which, however, fall under the same negative head of fallacies other than those of language.

```
              | In the language
              | (in the signs of thought)
              |
Fallacy -| |—In the Form.
         |—Outside the language -|
              | (in the thought itself) |
```

|
|—in the Matter.

§ 831. There are then three heads to which fallacies may be referred-namely, Formal Fallacies, Fallacies of Language, which are commonly known as Fallacies of Ambiguity, and, lastly, Material Fallacies.

§ 832. Aristotle himself only goes so far as the first step in the division of fallacies, being content to class them according as they are in the language or outside of it. After that he proceeds at once to enumerate the infimæ species under each of the two main heads. We shall presently imitate this procedure for reasons of expediency. For the whole phraseology of the subject is derived from Aristotle's treatise on Sophistical Refutations, and we must either keep to his method or break away from tradition altogether. Sufficient confusion has already arisen from retaining Aristotle's language while neglecting his meaning.

§ 833. Modern writers on logic do not approach fallacies from the same point of view as Aristotle. Their object is to discover the most fertile sources of error in solitary reasoning; his was to enumerate the various tricks of refutation which could be employed by a sophist in controversy. Aristotle's classification is an appendix to the Art of Dialectic.

§ 834. Another cause of confusion in this part of logic is the identification of Aristotle's two-fold division of fallacies, commonly known under the titles of In dictione and Extra diotionem, with the division into Logical and Material, which is based on quite a different principle.

§ 835. Aristotle's division perhaps allows an undue importance to language, in making that the principle of division, and so throwing formal and material fallacies under a common head. Accordingly another classification has been adopted, which concentrates attention from the first upon the process of thought, which ought certainly to be of primary importance in the eyes of the logician. This classification is as follows.

§ 836. Whenever in the course of our reasoning we are involved in error, either the conclusion follows from the premises or it does not. If it does not, the fault must lie in the process of reasoning, and we have then what is called a Logical Fallacy. If, on the other hand, the conclusion does follow from the premises, the fault must lie in the premises themselves, and we then have what is called a Material Fallacy. Sometimes, however, the conclusion will appear to follow from the premises until the meaning of the terms is examined, when it will be found that the appearance is deceptive owing to some ambiguity in the language. Such fallacies as these are, strictly speaking, non-logical, since the meaning of words is extraneous to the science which deals with thought. But they are called Semi-logical. Thus we arrive by a different road at the same three heads as before, namely, (1) Formal or Purely Logical Fallacies, (2) Semi-logical Fallacies or Fallacies of Ambiguity, (3) Material Fallacies.

§ 837. For the sake of distinctness we will place the two divisions side by side, before we proceed to enumerate the infimae species.

```
          |—In the language
          | (Fallacy of Ambiguity)
  Fallacy-|
          | |—In the Form.
          |—Outside the language -|
                    |
                    |—In the Matter.
              |—Formal or purely logical.
          |—Logical -|
  Fallacy-| |—Semi-logical
```

| (Fallacy of Ambiguity).

|—Material

838. Of one of these three heads, namely, formal fallacies, it is not necessary to say much, as they have been amply treated of in the preceding pages. A formal fallacy arises from the breach of any of the general rules of syllogism. Consequently it would be a formal fallacy to present as a syllogism anything which had more or less than two premises. Under the latter variety comes what is called 'a woman's reason,' which asserts upon its own evidence something which requires to be proved. Schoolboys also have been known to resort to this form of argument—'You're a fool.' 'Why?' 'Because you are.' When the conclusion thus merely reasserts one of the premises, the other must be either absent or irrelevant. If, on the other hand, there are more than two premises, either there is more than one syllogism or the superfluous premiss is no premiss at all, but a proposition irrelevant to the conclusion.

839. The remaining rules of the syllogism are more able to be broken than the first; so that the following scheme presents the varieties of formal fallacy which are commonly enumerated—

|—Four Terms.

Formal Fallacy-|—Undistributed Middle.

|—Illicit Process.

|—Negative Premisses and Conclusion.

§ 840. The Fallacy of Four Terms is a violation of the second of the general rules of syllogism (§ 582). Here is a palpable instance of it—

All men who write books are authors.

All educated men could write books.

∴ All educated men are authors.

Here the middle term is altered in the minor premiss to the destruction of the argument. The difference between the actual writing of books and the power to write them is precisely the difference between one who is an author and one who is not.

§ 841. Since a syllogism consists of three terms, each of which is used twice over, it would be possible to have an apparent syllogism with as many as six terms in it. The true name for the fallacy therefore is the Fallacy of More than Three Terms. But it is rare to find an attempted syllogism which has more than four terms in it, just as we are seldom tendered a line as an hexameter, which has more than seven feet.

§ 842. The Fallacies of Undistributed Middle and Illicit Process have been treated of under §§ 585, 586. The heading 'Negative Premisses and Conclusion' covers violations of the three general rules of syllogism relating to negative premisses (§§ 590-593). Here is an instance of the particular form of the fallacy which consists in the attempt to extract an affirmative conclusion out of two negative premisses—

All salmon are fish, for neither salmon nor fish belong to the class mammalia.

The accident of a conclusion being true often helps to conceal the fact that it is illegitimately arrived at. The formal fallacies which have just been enumerated find no place in Aristotle's division. The reason is plain. His object was to enumerate the various modes in which a sophist might snatch an apparent victory, whereas by openly violating any of the laws of syllogism a disputant would be simply courting defeat.

§ 843. We now revert to Aristotle's classification of fallacies, or rather of Modes of Refutation. We will take the species he enumerates in their order, and notice how modern usage has departed from the original meaning of the terms. Let it be borne in mind that, when the deception was not in the language, Aristotle did not trouble himself to determine whether it lay in the matter or in the form of thought.

§ 844. The following scheme presents the Aristotelian classification to the eye at a glance:—

```
| |—Equivocation.
| |—Amphiboly.
|—In the language -|—Composition.
| |—Division.
| |—Accent.
| |—Figure of Speech.
Modes of -|
Refutation. | |—Accident.
| |—A dicto secundum quid.
| |—Ignoratio Elenchi.
|—Outside the language -|—Consequent.
| |—Petitio Principii.
| |—Non causa pro causa.
| |—Many Questions.
```

[Footnote: for "In the language": The Greek is [Greek: para ten lexin], the exact meaning of which is; 'due to the statement.']

§ 845. The Fallacy of Equivocation [Greek: òmonumía] consists in an ambiguous use of any of the three terms of a syllogism. If, for instance, anyone were to argue thus—

No human being is made of paper,

All pages are human beings,

∴ No pages are made of paper—

the conclusion would appear paradoxical, if the minor term were there taken in a different sense from that which it bore in its proper premiss. This therefore would be an instance of the fallacy of Equivocal Minor.

§ 846. For a glaring instance of the fallacy of Equivocal Major, we may take the following—

No courageous creature flies,

The eagle is a courageous creature,

∴ The eagle does not fly—

the conclusion here becomes unsound only by the major being taken ambiguously.

§ 847. It is, however, to the middle term that an ambiguity most frequently attaches. In this case the fallacy of equivocation assumes the special name of the Fallacy of Ambiguous Middle. Take as an instance the following—

Faith is a moral virtue.

To believe in the Book of Mormon is faith.

∴ To believe in the Book of Mormon is a moral virtue.

Here the premisses singly might be granted; but the conclusion would probably be felt to be unsatisfactory. Nor is the reason far to seek. It is evident that belief in a book cannot be faith in any sense in which that quality can rightly be pronounced to be a moral virtue.

§ 848. The Fallacy of Amphiboly ([Greek: ámphibolía]) is an ambiguity attaching to the construction of a proposition rather than to the terms of which it is composed. One of Aristotle's examples is this—

[Greek: tò boúlesthai labeîn me toùs polemíous]

which may be interpreted to mean either 'the fact of my wishing to take the enemy,' or 'the fact of the enemies' wishing to take me.' The classical languages are especially liable to this fallacy owing to the oblique construction in which the accusative becomes subject to the verb. Thus in Latin we have the oracle given to Pyrrhus (though of course, if delivered at all, it must have been in Greek)—

Aio te, AEacida, Romanos vincere posse.

Pyrrhus the Romans shall, I say, subdue (Whately).

[Footnote: Cicero, De Divinatione, ii. § 116; Quintilian, Inst. Orat. vii 9, § 6.]

which Pyrrhus, as the story runs, interpreted to mean that he could conquer the Romans, whereas the oracle subsequently explained to him that the real meaning was that the Romans could conquer him. Similar to this, as Shakspeare makes the Duke of York point out, is the witch's prophecy in Henry VI (Second Part, Act i, sc. 4),

The duke yet lives that Henry shall depose.

An instance of amphiboly may be read on the walls of Windsor Castle—Hoc fecit Wykeham. The king mas incensed with the bishop for daring to record that he made the tower, but the latter adroitly replied that what he really meant to indicate was that the tower was the making of him. To the same head may be referred the famous sentence—'I will wear no clothes to distinguish me from my Christian brethren.'

§ 849. The Fallacy of Composition [Greek: diaíresis] is likewise a case of ambiguous construction. It consists, as expounded by Aristotle, in taking words together which ought to be taken separately, e.g.

'Is it possible for a man who is not writing to write?'
'Of course it is.'
'Then it is possible for a man to write without writing.'
And again—
'Can you carry this, that, and the other?' 'Yes.'
'Then you can carry this, that, and the other,'—
a fallacy against which horses would protest, if they could.

§ 850. It is doubtless this last example which has led to a convenient misuse of the term 'fallacy of composition' among modern writers, by whom it is defined to consist in arguing from the distributive to the collective use of a term.

§ 851. The Fallacy of Division ([Greek: diaíresis]), on the other hand, consists in taking words separately which ought to be taken together, e.g.

[Greek: ègó s' êteka doûlon ônt' èleúteron [Footnote: Evidently the original of the line in Terence's *Andria*, 37,—feci ex servo ut esses libertus mihi.],

where the separation of [Greek: doûlon] from [Greek: ôntra] would lead to an interpretation exactly contrary to what is intended.

And again—
[Greek: pentékont' àndrôn èkatòn lípe dîos Àchilleús],

where the separation of [Greek: àndrôn] from [Greek: èkatòn] leads to a ludicrous error.

Any reader whose youth may have been nourished on 'The Fairchild Family' may possibly recollect a sentence which ran somewhat on this wise—'Henry,' said Mr. Fairchild, 'is this true? Are you a thief and a liar too?' But I am afraid he will miss the keen delight which can be extracted at a certain age from turning the tables upon Mr. Fairchild thus—Henry said, 'Mr. Fairchild, is this true? Are *you* a thief and a liar too?'

§ 852. The fallacy of division has been accommodated by modern writers to the meaning which they have assigned to the fallacy of composition. So that by the 'fallacy of division' is now meant arguing from the collective to the distributive use of a term. Further, it is laid down that when the middle term is used distributively in the major premiss and collectively in the minor, we have the fallacy of composition; whereas, when the middle term is used collectively in the major premiss and distributively in the minor, we have the fallacy of division. Thus the first of the two examples appended would be composition and the second division.

(1) Two and three are odd and even.
Five is two and three.
∴. Five is odd and even.
(2) The Germans are an intellectual people.
Hans and Fritz are Germans.
∴. They are intellectual people.

§ 853. As the possibility of this sort of ambiguity is not confined to the middle term, it seems desirable to add that when either the major or minor term is used distributively in the premiss and collectively in the conclusion, we have the fallacy of composition, and in the converse case the fallacy of division. Here is an instance of the latter kind in which the minor term is at fault—

Anything over a hundredweight is too heavy to lift.
These sacks (collectively) are over a hundredweight.
.'. These sacks (distributively) are too heavy to lift.

§ 854. The ambiguity of the word 'all,' which has been before commented upon (§ 119), is a great assistance in the English language to the pair of fallacies just spoken of.

§ 835. The Fallacy of Accent ([Greek: prosodía]) is neither more nor less than a mistake in Greek accentuation. As an instance Aristotle gives Iliad xxiii. 328, where the ancient copies of Homer made nonsense of the words [Greek: tò mèn oú katapútetai ómbro] by writing [Greek: oû] with the circumflex in place of [Greek: oú] with the acute accent. [Footnote: This goes to show that the ancient Greeks did not distinguish in pronunciation between the rough and smooth breathing any more than their modern representatives.] Aristotle remarks that the fallacy is one which cannot easily occur in verbal argument, but rather in writing and poetry.

§ 856. Modern writers explain the fallacy of accent to be the mistake of laying the stress upon the wrong part of a sentence. Thus when the country parson reads out, 'Thou shall not bear false witness *against* thy neighbour,' with a strong emphasis upon the word 'against,' his ignorant audience leap [sic] to the conclusion that it is not amiss to tell lies provided they be in favour of one's neighbour.

§ 857. The Fallacy of Figure of Speech [Greek: tò schêma tês léxeos] results from any confusion of grammatical forms, as between the different genders of nouns or the different voices of verbs, or their use as transitive or intransitive, e.g. [Greek: úgiaínein] has the same grammatical form as [Greek: témnein] or [Greek: oìkodomeîn], but the former is intransitive, while the latter are transitive. A sophism of this kind is put into the mouth of Socrates by Aristophanes in the Clouds (670-80). The philosopher is there represented as arguing that [Greek: kápdopos] must be masculine because [Greek: Kleónumos] is. On the surface this is connected with language, but it is essentially a fallacy of false analogy.

§ 858. To this head may be referred what is known as the Fallacy of Paronymous Terms. This is a species of equivocation which consists in slipping from the use of one part of speech to that of another, which is derived from the same source, but has a different meaning. Thus this fallacy would be committed if, starting from the fact that there is a certain probability that a hand at whist will consist of thirteen trumps, one were to proceed to argue that it was probable, or that he had proved it.

§ 859. We turn now to the tricks of refutation which lie outside the language, whether the deception be due to the assumption of a false premiss or to some unsoundness in the reasoning.   •

§ 860. The first on the list is the Fallacy of Accident ([Greek: tò sumbebekós]). This fallacy consists in confounding an essential with an accidental difference, which is not allowable, since many things are the same in essence, while they differ in accidents. Here is the sort of example that Aristotle gives—

'Is Plato different from Socrates ?' 'Yes.' 'Is Socrates a man ?'
'Yes.' 'Then Plato is different from man.'

To this we answer—No: the difference of accidents between Plato and Socrates does not go so deep as to affect the underlying essence. To put the thing more plainly, the fallacy lies in assuming that whatever is different from a given subject must be different from it in all respects, so that it is impossible for them to have a common predicate. Here Socrates and Plato, though different from one another, are not so different but that they have the common predicate 'man.' The attempt to prove that they have not involves an illicit process of the major.

§ 861. The next fallacy suffers from the want of a convenient name. It is called by Aristotle [Greek: tò áplos tóde ê pê légestai kaì mè kupíos] or, more briefly, [Greek: tò áplôs ê mé], or [Greek: tò pê kaí áplôs], and by the Latin writers 'Fallacia a dicto secundum quid ad dictum simpliciter.' It consists in taking what is said in a particular respect as though it held true without any restriction, e.g., that because the nonexistent ([Greek: tò mè ôn]) is a matter of opinion, that therefore the non-existent is, or again that because the existent ([Greek: tò ôn]) is not a man, that therefore the existent is not. Or again, if an Indian, who as a whole is black, has white teeth, we should be committing this species of fallacy in declaring him to be both white and not-white. For he is only white in a certain respect ([Greek: pê]), but not absolutely ([Greek: àplôs]). More difficulty, says Aristotle, may arise when opposite qualities exist in a thing in about an equal degree. When, for instance, a thing is half white and half black, are we to say that it is white or black? This question the philosopher propounds, but does not answer. The force of it lies in the implied attack on the Law of Contradiction. It would seem in such a case that a thing may be both white and not-white at the same time. The fact is—so subtle are the ambiguities of language—that even such a question as 'Is a thing white or not-white?' straightforward, as it seems, is not really a fair one. We are entitled sometimes to take the bull by the horns, and answer with the adventurous interlocutor in one of Plato's dialogues—'Both and neither.' It may be both in a certain respect, and yet neither absolutely.

§ 862. The same sort of difficulties attach to the Law of Excluded Middle, and may be met in the same way. It might, for instance, be urged that it could not be said with truth of the statue seen by Nebuchadnezzar in his dream either that it was made of gold or that it was not made of gold: but the apparent plausibility of the objection would be due merely to the ambiguity of language. It is not true, on the one hand, that it was made of gold (in the sense of being composed entirely of that metal); and it is not true, on the other, that it was not made of gold (in the sense of no gold at all entering into its composition). But let the ambiguous proposition be split up into its two meanings, and the stringency of the Law of Excluded Middle will at once appear—

(1) It must either have been composed entirely of gold or not.

(2) Either gold must have entered into its composition or not.

§ 863. By some writers this fallacy is treated as the converse of the last, the fallacy of accident being assimilated to it under the title of the 'Fallacia a dicto simpliciter ad dictum secundum quid.' In this sense the two fallacies may be defined thus.

The Fallacy of Accident consists in assuming that what holds true as a general rule will hold true under some special circumstances which may entirely alter the case. The Converse Fallacy of Accident consists in assuming that what holds true under some special circumstances must hold true as a general rule.

The man who, acting on the assumption that alcohol is a poison, refuses to take it when he is ordered to do so by the doctor, is guilty of the fallacy of accident; the man who, having had it prescribed for him when he was ill, continues to take it morning, noon, and night, commits the converse fallacy.

§ 864. There ought to be added a third head to cover the fallacy of arguing from one special case to another.

§ 865. The next fallacy is Ignoratio Elenchi [Greek: èlégchou âgnoia]. This fallacy arises when by reasoning valid in itself one establishes a conclusion other than what is required to upset the adversary's assertion. It is due to an inadequate conception of the true nature of refutation. Aristotle therefore is at the pains to define refutation at full length, thus—

'A refutation [Greek: êlegchos] is the denial of one and the same—not name, but thing, and by means, not of a synonymous term, but of the same term, as a necessary consequence from the data, without assumption of the point originally at issue, in the same respect, and in the same relation, and in the same way, and at the same time.'

The ELENCHUS then is the exact contradictory of the opponent's assertion under the terms of the law of contradiction. To establish by a syllogism, or series of syllogisms, any other proposition, however slightly different, is to commit this fallacy. Even if the substance of the contradiction be established, it is not enough unless the identical words of the opponent are employed in the contradictory. Thus if his thesis asserts or denies something about [Greek: lópion], it is not enough for you to prove the contradictory with regard to [Greek: imátion]. There will be need of a further question and answer to identify the two, though they are admittedly synonymous. Such was the rigour with which the rules of the game of dialectic were enforced among the Greeks!

§ 866. Under the head of Ignoratio Elenchi it has become usual to speak of various forme of argument which have been labelled by the Latin writers under such names as 'argumentum ad hominem,' 'ad populum,' 'ad verecundiam,' 'ad ignorantiam,' 'ad baculum'—all of them opposed to the 'argumentum ad rem' or 'ad judicium.'

§ 867. By the 'argumentum ad hominem' was perhaps meant a piece of reasoning which availed to silence a particular person, without touching the truth of the question. Thus a quotation from Scripture is sufficient to stop the mouth of a believer in the inspiration of the Bible. Hume's Essay on Miracles is a noteworthy instance of the 'argumentum ad hominem' in this sense of the term. He insists strongly on the evidence for certain miracles which he knew that the prejudices of his hearers would prevent their ever accepting, and then asks triumphantly if these miracles, which are declared to have taken place in an enlightened age in the full glare of publicity, are palpably imposture, what credence can be attached to accounts of extraordinary occurrences of remote antiquity, and connected with an obscure corner of the globe? The 'argumentum ad judicium' would take miracles as a whole, and endeavour to sift the amount of truth which may lie in the accounts we have of them in every age. [Footnote: On this subject see the author's *Attempts at Truth* (Trubner & Co.), pp. 46-59.]

§ 868. In ordinary discourse at the present day the term 'argumentum ad hominem' is used for the form of irrelevancy which consists in attacking the character of the opponent instead of combating his arguments, as illustrated in the well-known instructions to a barrister—'No case: abuse the plaintiff's attorney.'

§ 869. The 'argumentum ad populum' consists in an appeal to the passions of one's audience. An appeal to passion, or to give it a less question-begging name, to feeling, is not necessarily amiss. The heart of man is the instrument upon which the rhetorician plays, and he has to answer for the harmony or the discord that comes of his performance.

§ 870. The 'argumentum ad verecundiam' is an appeal to the feeling of reverence or shame. It is an argument much used by the old to the young and by Conservatives to Radicals.

§ 871. The 'argumentum ad ignorantiam' consists simply in trading on the ignorance of the person addressed, so that it covers any kind of fallacy that is likely to prove effective with the hearer.

§ 872. The 'argumentum ad baculum' is unquestionably a form of irrelevancy. To knock a man down when he differs from you in opinion may prove your strength, but hardly your logic.

A sub-variety of this form of irrelevancy was exhibited lately at a socialist lecture in Oxford, at which an undergraduate, unable or unwilling to meet the arguments of the speaker, uncorked a bottle, which had the effect of instantaneously dispersing the audience. This might be set down as the 'argumentum ad nasum.'

§ 873. We now come to the Fallacy of the Consequent, a term which has been more hopelessly abused than any. What Aristotle meant by it was simply the assertion of the consequent in a conjunctive proposition, which amounts to the same thing as the simple conversion of A (§ 489), and is a fallacy of distribution. Aristotle's example is this—

If it has rained, the ground is wet.

.'. If the ground is wet, it has rained.

This fallacy, he tells us, is often employed in rhetoric in dealing with presumptive evidence. Thus a speaker, wanting to prove that a man is an adulterer, will argue that he is a showy dresser, and has been seen about at nights. Both these things however may be the case, and yet the charge not be true.

§ 874. The Fallacy of Petitio or Assumptio Principii [Greek: tò èn àrchê aìteîstai or lambánein] to which we now come, consists in an unfair assumption of the point at issue. The word [Greek: aìteîstai], in Aristotle's name for it points to the Greek method of dialectic by means of question and answer. This fact is rather disguised by the mysterious phrase 'begging the question.' The fallacy would be committed when you asked your opponent to grant, overtly or covertly, the very proposition originally propounded for discussion.

§ 875. As the question of the precise nature of this fallacy is of some importance we will take the words of Aristotle himself (Top. viii. 13. §§ 2, 3). 'People seem to beg the question in five ways. First and most glaringly, when one takes for granted the very thing that has to be proved. This by itself does not readily escape detection, but in the case of "synonyms," that is, where the name and the definition have the same meaning, it does so more easily. [Footnote: Some light is thrown upon this obscure passage by a comparison with Cat. I. § 3, where 'synonym' is defined. To take the word here in its later and modern sense affords an easy interpretation, which is countenanced by Alexander Aphrodisiensis, but it is flat against the usage of Aristotle, who elsewhere gives the name 'synonym,' not to two names for the same thing, but to two things going under the same name. See Trendelenberg on the passage.]

Secondly, when one assumes universally that which has to be proved in particular, as, if a man undertaking to prove that there is one science of contraries, were to assume that there is one science of opposites generally. For he seems to be taking for granted along with several other things what he ought to have proved by itself.

Thirdly, when one assumes the particulars where the universal has to be proved; for in so doing a man is taking for granted separately what he was bound to prove along with several other things. Again, when one assumes the question at issue by splitting it up, for instance, if, when the point to be proved is that the art of medicine deals with health and disease, one were to take each by itself for granted.

Lastly, if one were to take for granted one of a pair of necessary consequences, as that the side is incommensurable with the diagonal, when it is required to prove that the diagonal is incommensurable with the side.'

§ 876. To sum up briefly, we may beg the question in five ways—

(1) By simply asking the opponent to grant the point which requires to be proved;

(2) by asking him to grant some more general truth which involves it;

(3) by asking him to grant the particular truths which it involves;

(4) by asking him to grant the component parts of it in detail;

(5) by asking him to grant a necessary consequence of it.

§ 877. The first of these five ways, namely, that of begging the question straight off, lands us in the formal fallacy already spoken of (§ 838), which violates the first of the general rules of syllogism, inasmuch as a conclusion is derived from a single premiss, to wit, itself.

§ 878. The second, strange to say, gives us a sound syllogism in Barbara, a fact which countenances the blasphemers of the syllogism in the charge they bring against it of containing in itself a petitio principii. Certainly Aristotle's expression might have been more guarded. But it is clear that his quarrel is with the matter, not with the form in such an argument. The fallacy consists in assuming a proposition which the opponent would be entitled to deny. Elsewhere Aristotle tells us that the fallacy arises when a truth not evident by its own light is taken to be so. [Footnote: [Greek: Ôtan tò mè dí aùtoû gnostòn dí aùtoû tis èpicheiraê deiknúnai, tót' aìteîtai tò èx àrchês.]. Anal. Pr. II. 16. § I ad fin.]

§ 879. The third gives us an inductio per enumerationem simplicem, a mode of argument which would of course be unfair as against an opponent who was denying the universal.

§ 880. The fourth is a more prolix form of the first.

§ 881. The fifth rests on Immediate Inference by Relation (§ 534).

§ 882. Under the head of petitio principii comes the fallacy of Arguing in a Circle, which is incidental to a train of reasoning. In its most compressed form it may be represented thus—

   (1) B is A.
   C is B.
   .'. C is A.
   (2) C is A.
   B is C.
   .'. B is A.

§ 883. The Fallacy of Non causa pro causa ([Greek: tò mè aîtion] or [Greek: aîtoin]) is another, the name of which has led to a complete misinterpretation. It consists in importing a contradiction into the discussion, and then fathering it on the position controverted. Such arguments, says Aristotle, often impose upon the users of them themselves. The instance he gives is too recondite to be of general interest.

§ 884. Lastly, the Fallacy of Many Questions ([Greek: tò tà déo èrotémata ên poieîn]) is a deceptive form of interrogation, when a single answer is demanded to what is not really a single question. In dialectical discussions the respondent was limited to a simple 'yes' or 'no'; and in this fallacy the question is so framed as that either answer would seem to imply the acceptance of a proposition which would be repudiated. The old stock instance will do as well as another—'Come now, sir, answer "yes" or "no." Have you left off beating your mother yet?' Either answer leads to an apparent admission of impiety.

A late Senior Proctor once enraged a man at a fair with this form of fallacy. The man was exhibiting a blue horse; and the distinguished stranger asked him—'With what did you paint your horse?'

EXERCISES.

These exercises should be supplemented by direct questions upon the text, which it is easy for the student or the teacher to supply for himself.

PART I.

CHAPTER I.

Classify the following words according as they are categorematic, syncategorematic or acategorematic;—

come peradventure why through inordinately pshaw therefore circumspect puss grand inasmuch stop touch sameness back cage disconsolate candle.

CHAPTER II.

Classify the following things according as they are substances, qualities or relations;—
God likeness weight blueness grass imposition ocean introduction thinness man air spirit Socrates raillery heat mortality plum fire.

CHAPTER III.

1. Give six instances each of-attribute, abstract, singular, privative, equivocal and relative terms.

2. Select from the following list of words such as are terms, and state whether they are (1) abstract or concrete, (2) singular or common, (3) univocal or equivocal:—
van table however enter decidedly tiresome very butt Solomon infection bluff Czar short although Caesarism distance elderly Nihilist.

3. Which of the following words are abstract terms?—
quadruped event through hate desirability thorough fact expressly thoroughness faction wish light inconvenient will garden inconvenience volition grind.

4. Refer the following terms to their proper place under each of the divisions in the scheme:—
horse husband London free lump empty liberty rational capital impotent reason Capitol impetuosity irrationality grave impulsive double calf.

5. Give six instances each of proper names and designations.

6. Give six instances each of connotative and non-connotative terms.

7. Give the extension and intension of—
sermon animal sky clock square gold sport fish element bird student fluid art river line gas servant language

CHAPTER IV.

Arrange the following terms in order of extension—carnivorous, thing, matter, mammal, organism, vertebrate, cat, substance, animal.
* * * * *

PART II.

CHAPTER I.

Give a name to each of the following sentences:—
(1) Oh, that I had wings like a dove!
(2) The more, the merrier.
(3) Come rest in this bosom, my own stricken deer.
(4) Is there balm in Gilead?
(5) Hearts may be trumps.

CHAPTER II.

Analyse the following propositions into subject, copula and predicate:—
(1) He being dead yet speaketh.
(2) There are foolish politicians.
(3) Little does he care.
(4) There is a land of pure delight.
(5) All's well that ends well.
(6) Sweet is the breath of morn.
(7) Now it came to pass that the beggar died.
(8) Who runs may read.
(9) Great is Diana of the Ephesians.
(10) Such things are.
(11) Not more than others I deserve.
(12) The day will come when Ilium's towers shall perish.

CHAPTER III.

1. Express in logical form, affixing the proper symbol:—
(1) Some swans are not white.
(2) All things are possible to them that believe.
(3) No politicians are unprincipled.
(4) Some stones float on water.
(5) The snow has melted.
(6) Eggs are edible.
(7) All kings are not wise.
(8) Moths are not butterflies.
(9) Some men are born great.
(10) Not all who are called are chosen.
(11) It is not good for man to be alone.
(12) Men of talents have been known to fail in life.
(13) 'Tis none but a madman would throw about fire.
(14) Every bullet does not kill.
(15) Amongst Unionists are Whigs.
(16) Not all truths are to be told.
(17) Not all your efforts can save him.
(18) The whale is a mammal.
(19) Cotton is grown in Cyprus.
(20) An honest man's the noblest work of God.
(21) No news is good news.
(22) No friends are like old friends.
(23) Only the ignorant affect to despise knowledge.
(24) All that trust in Him shall not be ashamed.
(25) All is not gold that glitters.
(26) The sun shines upon the evil and upon the good.
(27) Not to go on is to go back.
(28) The king, minister, and general are a pretty trio.
(29) Amongst dogs are hounds.
(30) A fool is not always wrong.
(31) Alexander was magnanimous.

(32) Food is necessary to life.
(33) There are three things to be considered,
(34) By penitence the Eternal's wrath's appeased.
(35) Money is the miser's end.
(36) Few men succeed in life.
(37) All is lost, save honour.
(38) It is mean to hit a man when he is down.
(39) Nothing but coolness could have saved him.
(40) Books are generally useful.
(41) He envies others' virtue who has none himself.
(42) Thankless are all such offices.
(43) Only doctors understand this subject.
(44) All her guesses but two were correct.
(45) All the men were twelve.
(46) Gossip is seldom charitable.

2. Give six examples of indefinite propositions, and then quantify them according to their matter.

3. Compose three propositions of each of the following kinds:—
(1) with common terms for subjects;
(2) with abstract terms for subjects;
(3) with singular terms for predicates;
(4) with collective terms for predicates;
(5) with attributives in their subjects;
(6) with abstract terms for predicates.

CHAPTER IV.

1. Point out what terms are distributed or undistributed in the following propositions:—
(1) The Chinese are industrious.
(2) The angle in a semi-circle is a right angle.
(3) Not one of the crew survived.
(4) The weather is sometimes not propitious.
The same exercise may be performed upon any of the propositions in the preceding list.
2. Prove that in a negative proposition the predicate must be distributed.

CHAPTER V.

Affix its proper symbol to each of the following propositions:—
(1) No lover he who is not always fond.
(2) There are Irishmen and Irishmen.
    (3) Men only disagree, Of creatures rational.
(4) Some wise men are poor.
(5) No Popes are some fallible beings.
(6) Some step-mothers are not unjust.
(7) The most original of the Roman poets was Lucretius.
    (8) Some of the immediate inferences are all the forms of conversion.

CHAPTER VI.

1. Give six examples of terms standing one to another as genus to species.

2. To which of the heads of predicables would you refer the following statements? And why?

(1) A circle is the largest space that can be contained by one line.

(2) All the angles of a square are right angles.

(3) Man alone among animals possesses the faculty of laughter.

(4) Some fungi are poisonous.

(5) Most natives of Africa are negroes.

(6) All democracies are governments.

(7) Queen Anne is dead.

CHAPTER VII.

1. Define the following terms—

Sun inn-keeper tea-pot hope anger virtue bread diplomacy milk carpet man death sincerity telescope mountain poverty Senate novel.

2. Define the following terms as used in Political Economy—

Commodity barter value wealth land price money labour rent interest capital wages credit demand profits.

3. Criticise the following as definitions—

(1) Noon is the time when the shadows of bodies are shortest.

(2) Grammar is the science of language.

(3) Grammar is a branch of philology.

(4) Grammar is the art of speaking and writing a language with propriety.

(5) Virtue is acting virtuously.

(6) Virtue is that line of conduct which tends to produce happiness.

(7) A dog is an animal of the canine species.

(8) Logic is the art of reasoning.

(9) Logic is the science of the investigation of truth by means of evidence.

(10) Music is an expensive noise.

(11) The sun is the centre of the solar system.

(12) The sun is the brightest of those heavenly bodies that move round the earth.

(13) Rust is the red desquamation of old iron.

(14) Caviare is a kind of food.

(15) Life is the opposite of death.

(16) Man is a featherless biped.

(17) Man is a rational biped.

(18) A gentleman is a person who has no visible means of subsistence.

(19) Fame is a fancied life in others' breath.

(20) A fault is a quality productive of evil or inconvenience.

(21) An oligarchy is the supremacy of the rich in a state.

(22) A citizen is one who is qualified to exercise deliberative and judicial functions.

(23) Length is that dimension of a solid which would be measured by the longest line.

(24) An eccentricity is a peculiar idiosyncrasy.

(25) Deliberation is that species of investigation which is concerned with matters of action.

(26) Memory is that which helps us to forget.

(27) Politeness is the oil that lubricates the wheels of society.

(28) An acute-angled triangle is one which has an acute angle.

(29) A cause is that without which something would not be.

(30) A cause is the invariable antecedent of a phenomenon.

(31) Necessity is the mother of invention.

(32) Peace is the absence of war.

(33) A net is a collection of holes strung together.

(34) Prudence is the ballast of the moral vessel.

(35) A circle is a plane figure contained by one line.

(36) Superstition is a tendency to look for constancy where constancy is not to be expected.

(37) Bread is the staff of life.

(38) An attributive is a term which cannot stand as a subject.

(39) Life is bottled sunshine.

(40) Eloquence is the power of influencing the feelings by speech or writing.

(41) A tombstone is a monument erected over a grave in memory of the dead.

(42) Whiteness is the property or power of exciting the sensation of white.

(43) Figure is the limit of a solid.

(44) An archdeacon is one who exercises archidiaconal functions.

(45) Humour is thinking in jest while feeling in earnest.

CHAPTER VIII.

1. Divide the following terms—

Soldier end book church good oration apple cause school ship government letter vehicle science verse.

2. Divide the following terms as used in Political Economy—

Requisites of production, labour, consumption, stock, wealth, capital.

3. Criticise the following as divisions—

(1) Great Britain into England, Scotland, Wales, and Ireland.

(2) Pictures into sacred, historical, landscape, and mythological.

(3) Vertebrate animals into quadrupeds, birds, fishes, and reptiles.

(4) Plant into stem, root, and branches.

(5) Ship into frigate, brig, schooner, and merchant-man.

(6) Books into octavo, quarto, green, and blue.

(7) Figure into curvilinear and rectilinear.

(8) Ends into those which are ends only, means and ends, and means only.

(9) Church into Gothic, episcopal, high, and low.

(10) Sciences into physical, moral, metaphysical, and medical.

(11) Library into public and private.

(12) Horses into race-horses, hunters, hacks, thoroughbreds, ponies, and mules.

4. Define and divide—

Meat, money, virtue, triangle;

and give, as far as possible, a property and accident of each.

PART III.

CHAPTERS I-III.

1. What kind of influence have we here?

The author of the Iliad was unacquainted with writing.

Homer was the author of the Iliad.

.'. Homer was unacquainted with writing.

2. Give the logical opposites of the following propositions—

(1) Knowledge is never useless.

(2) All Europeans are civilised.

(3) Some monks are not illiterate.

(4) Happy is the man that findeth wisdom.

(5) No material substances are devoid of weight.

(6) Every mistake is not culpable.

(7) Some Irishmen are phlegmatic.

3. Granting the truth of the following propositions, what other propositions can be inferred by opposition to be true or false?

(1) Men of science are often mistaken.

(2) He can't be wrong, whose life is in the right.

(3) Sir Walter Scott was the author of Waverley.

(4) The soul that sinneth it shall die.

(5) All women are not vain.

4. Granting the falsity of the following propositions, what other propositions can be inferred by opposition to be true or false?—

(1) Some men are not mortal.

(2) Air has no weight.

(3) All actors are improper characters.

(4) None but dead languages are worth studying.

(5) Some elements are compound.

CHAPTER IV.

1. Give, as far as possible, the logical converse of each of the following propositions—

(1) Energy commands success.

(2) Mortals cannot be happy.

(3) There are mistakes which are criminal.

(4) All's well that ends well.

(5) Envious men are disliked.

(6) A term is a kind of word or collection of words.

(7) Some Frenchmen are not vivacious.

(8) All things in heaven and earth were hateful to him.

(9) The square of three is nine.

(10) All cannot receive this saying.

(11) P struck Q.

(12) Amas.

2. 'More things may be contained in my philosophy than exist in heaven or earth: but the converse proposition is by no means true.' Is the term converse here used in its logical meaning?

CHAPTER V.

Permute the following propositions—

(1) All just acts are expedient.

(2) No display of passion is politic.

(3) Some clever people are not prudent.

(4) Some philosophers have been slaves.

The same exercise may be performed upon any of the propositions in the preceding lists.

CHAPTER VI.

1. Give the converse by negation of—

(1) All women are lovely.

(2) Some statesmen are not practical.

(3) All lawyers are honest.

(4) All doctors are skilful.

(5) Some men are not rational.

2. Give the contrapositive of—

(1) All solid substances are material.

(2) All the men who do not row play cricket.

(3) All impeccable beings are other than human,

(4) Some prejudiced persons are not dishonest.

3. Prove indirectly the truth of the contrapositive of 'All A is B.'

4. Criticise the following as immediate inferences—

(1) All wise men are modest.

.'. No immodest men are wise.

(2) Some German students are not industrious.

.'. Some industrious students are not Germans.

(3) Absolute difference excludes all likeness.

.'. Any likeness is a proof of sameness.

(4) None but the brave deserve the fair.

.'. All brave men deserve the fair.

(5) All discontented men are unhappy.

.'. No contented men are unhappy.

(6) Books being a source of instruction, our knowledge must come from our libraries.

(7) All Jews are Semitic.

.'. Some non-Semitic people are not Jews.

5. Show by what kind of inference each of the subjoined propositions follows from All discontented men are unhappy.

(1) All happy men are contented.

(2) Some discontented men are unhappy.

(3) Some contented men are happy.

(4) Some unhappy men are not contented.

(5) No discontented men are happy.

(6) Some happy men are contented.

(7) Some contented men are not unhappy.

(8) Some unhappy men are discontented.

(9) No happy men are discontented.

(10) Some discontented men are not happy.

(11) Some happy men are not discontented.

(12) None but unhappy men are discontented.

From how many of these propositions can the original one be derived? And why not from all?

CHAPTER VII.

What kind of inference have we here?—
(1) None but the ignorant despise knowledge.
∴ No wise man despises knowledge.
(2) A is superior to B.
∴ B is inferior to A.

CHAPTER VIII.

Fill up the following enthymemes, mentioning to which order they belong, and state which of them are expressed in problematic form—
(1) I am fond of music: for I always like a comic song.
(2) All men are born to suffering, and therefore you must expect your share.
(3) Job must have committed some secret sins: for he fell into dreadful misfortunes.
(4) Latin was the language of the Vestals, and therefore no lady need be ashamed of speaking it.
(5) None but physicians came to the meeting. There were therefore no nurses there.
(6) The human soul extends through the whole body, for it is found in every member.
(7) No traitor can be trusted, and you are a traitor.
(8) Whatever has no parts does not perish by the dissolution of its parts. Therefore the soul of man is imperishable.
Is the suppressed premiss in any case disputable on material grounds?

CHAPTERS IX-XVIII.

Refer the following arguments to their proper mood and figure, or show what rules of syllogism they violate.
(1) No miser is a true friend, for he does not assist his friend with his purse.
(2) Governments are good which promote prosperity.
The government of Burmah does not promote prosperity.
∴ It is not a good government.
(3) Land is not property.
Land produces barley.
∴ Beer is intoxicating.
(4) Nothing is property but that which is the product of man's hand.
The horse is not the product of man's hand.
∴ The horse is not property.
(5) Some Europeans at least are not Aryans, because the Finns are not.

(6) Saturn is visible from the earth, and the moon is visible from the earth. Therefore the moon is visible from Saturn.

(7) Some men of self-command are poor, and therefore some noble characters are poor.

(8) Sparing the rod spoils the child: so John will turn out very good, for his mother beats him every day.

(9) Some effects of labour are not painful, since every virtue is an effect of labour.

(10) The courageous are confident and the experienced are confident. Therefore the experienced are courageous.

(11) No tale-bearer is to be trusted, and therefore no great talker is to be trusted, for all tale-bearers are great talkers.

(12) Socrates was wise, and wise men alone are happy: therefore Socrates was happy.

II.

1. From the major 'No matter thinks' draw, by supplying the minor, the following conclusions—

(1) Some part of man does not think.

(2) The soul of man is not matter.

(3) Some part of man is not matter.

(4) Some substance does not think.

Name the figured mood into which each syllogism falls.

2. Construct syllogisms in the following moods and figures, stating whether they are valid or invalid, and giving your reasons in each case—

AEE in the first figure; EAO in the second; IAI in the third; AII in the fourth.

3. Prove that 'Brass is not a metal,' using as your middle term 'compound body.'

4. Construct syllogisms to prove or disprove—

(1) Some taxes are necessary.

(2) No men are free.

(3) Laws are salutary.

5. Prove by a syllogism in Bokardo that 'Some Socialists are not unselfish,' and reduce your syllogism directly and indirectly.

6. Prove the following propositions in the second figure, and reduce the syllogisms you use to the first—

(1) All negroes are not averse to education.

(2) Only murderers should be hanged.

7. Prove in Baroko and also in Ferio that 'Some Irishmen are not Celts.'

8. Construct in words the same syllogism in all the four figures.

9. Invent instances to show that false premisses may give true conclusions.

III.

1. What moods are peculiar to the first, second, and third figures respectively?

2. What moods are common to all the figures?

3. Why can there be no subaltern moods in the third figure?

4. What is the only kind of conclusion that can be drawn in all the figures?

5. Show that IEO violates the special rules of all the figures.

6. In what figures is AEE valid?

7. Show that AEO is superfluous in any figure.

8. Prove that O cannot be a premiss in the first figure, nor a minor premiss anywhere but in the second.

9. Show that in the first figure the conclusion must have the quality of the major premiss and the quantity of the minor.

10. Why do the premisses EA yield a universal conclusion in the first two figures and only a particular one in the last two?

11. Show that AAI is the only mood in the fourth figure in which it is possible for the major term to be distributed in the premiss and undistributed in the conclusion.

12. Why are the premisses of Fesapo and Fresison not transposed in reduction like those of the other moods of the fourth figure?

IV.

1. Why is it sufficient to distribute the middle term once only?

2. Prove that from two affirmative premisses you cannot get a negative conclusion.

3. Prove that there must be at least one more term distributed in the premisses than in the conclusion.

4. Prove that the number of distributed terms in the premisses cannot exceed those in the conclusion by more than two.

5. Prove that the number of undistributed terms in the premisses cannot exceed those in the conclusion by more than one.

6. Prove that wherever the minor premiss is negative, the major must be universal.

7. Prove that wherever the minor term is distributed, the major premiss must be universal.

8. If the middle term be twice distributed, what mood and figure are possible?

9. If the major term of a syllogism be the predicate of the major premiss, what do we know about the minor premiss?

10. When the middle term is distributed in both premisses, what must be the quantity of the conclusion?

11. Prove that if the conclusion be universal, the middle term can only be distributed once in the premisses.

12. Show how it is sometimes possible to draw three different conclusions from the same premisses.

CHAPTER XIX.

1. Convert the following propositions—

(1) If a man is wise, he is humble.

(2) Where there is sincerity there is no affectation.

(3) When night-dogs run, all sorts of deer are chased.

(4) The nearer the Church, the further from God.

(5) If there were no void, all would be solid.

(6) Not to go on is sometimes to go back.

2. Express in a single proposition—

If he was divine, he was not covetous; and if he was covetous, he was not divine.

3. Exhibit the exact logical relation to one another of the following pairs of propositions—

(1) If the conclusion be false, the premisses are false. If the conclusion be true, the premisses are not necessarily true.

(2) If one premiss be negative, the conclusion must be negative.

If the conclusion be negative, one of the premisses must be negative.

(3) The truth of the universal involves the truth of the particular.

The falsity of the particular involves the falsity of the universal.

(4) From the truth of the particular no conclusion follows as to the universal.

From the falsity of the universal no conclusion follows as to the particular.

(5) If the conclusion in the fourth figure be negative, the major premiss must be universal.

If the major premiss in the fourth figure be particular, the conclusion must be affirmative.

(6) If both premisses be affirmative, the conclusion must be affirmative.

If the conclusion be negative, one of the premisses must be negative.

4. 'The Method of Agreement stands on the ground that whatever circumstance can be eliminated is not connected with the phenomenon by any law; the Method of Difference stands on the ground that whatever circumstance cannot be eliminated is connected with the phenomenon by a law.' Do these two principles imply one another?

CHAPTERS XX-XXVIII.

1. Fill up the following enthymemes, and state the exact nature of the resulting syllogism—

(1) If Livy is a faultless historian, we must believe all that he tells us; but that it is impossible to do.

(2) If they stay abroad, the wife will die; while the husband's lungs will not stand the English climate. It is to be feared therefore that one must fall a victim.

(3) He is either very good, very bad, or commonplace. But he is not very good.

(4) Either a slave is capable of virtue or he is not.

∴ Either he ought not to be a slave or he is not a man.

(5) Does not his feebleness of character indicate either a bad training or a natural imbecility?

(6) Those who ask shan't have; those who don't ask don't want.

(7) If a man be mad, he deviates from the common standard of intellect.

∴ If all men be alike mad, no one is mad.

(8) 'I cannot dig; to beg I am ashamed.'

2. 'The infinite divisibility of space implies that of time. If the latter therefore be impossible, the former must be equally so.' Formulate this argument as an immediate inference.

3. Examine the following arguments—

(1) If we have a dusty spring, there is always a good wheat harvest. We shall therefore have a poor harvest this year, for the spring has not been dusty.

(2) Virtues are either feelings, capacities, or states; and as they are neither feelings nor capacities, they must be states.

(3) Everything must be either just or unjust.
Justice is a thing, and is not unjust.
.'. Justice is just.
  Similarly justice is holy.
But the virtues of knowledge, justice, courage, temperance, and
holiness were declared to be different from one another.
.'. Justice is unholy and holiness unjust.

CHAPTER XXIX.

Formulate the following trains of reasoning, resolve them into their component parts, and point out any violations of the rules of syllogism which they may contain—

(1) No Church Institutions are useful; for they teach religious matters, not business matters, which latter are useful, being profitable.

(2) Mr. Darwin long ago taught us that the clover crop is dependent on the number of maiden ladies in the district. For the ladies keep cats, and the cats destroy the field-mice, which prey on the bees, which, in their turn, are all-important agents in the fertilisation of the clover flowers.

(3) Athletic games are duties; for whatever is necessary to health is a duty, and exercise is necessary to health, and these games are exercise.

(4) The iron-trade leads to the improvement of a new country; for furnaces require to be fed with fuel, which causes land to be cleared.

(5) 'Is stone a body?' 'Yes.' 'Well, is not an animal a body?'
'Yes,' 'And are you an animal?' 'It seems so.' 'Then you are a
stone, being an animal.'

(6) If A is B, C is D.
If E is F, G is H.
But if A is B, E is F.
.'. If C is D, G is sometimes H.

(7) The soul is not matter.
My arm is not myself.

(8) Honesty deserves reward and a negro is a fellow-creature. Therefore an honest negro is a fellow-creature deserving of reward.

CHAPTER XXX.

1. Point out any ambiguities which underlie the following propositions—

(1) Every one who has read the book in French will recommend those who have not to read it in English.

(2) I will not do this because he did it.

(3) These are all my books.

(4) By an old statute of the date of Edward III it was accorded 'that Parliament should be holden every year once or more often if need be.'

(5) They found Mary and Joseph and the babe lying in a manger.

(6) The king and his minister are feeble and unscrupulous.

(7) Heres meus uxori meae triginta pondo vasorum argenteorum dato, quae volet.

2. Examine the following arguments, formulating them when sound, and referring them, when unsound, to the proper head of fallacy—

(1) We know that thou art a teacher come from God; for no man can do these signs that thou doest, except God be with him. S. John iii. 2.

(2) 'Sir Walter Scott's novels have ceased to be popular.' 'Well, that's only because nobody reads them.'

(3) What we produce is property.
  The sheriff produces a prisoner.
  .'. A prisoner is property.

(4) As all metals are not necessarily solid, we may expect some metals to be liquid.

(5) Moses was the son of Pharaoh's daughter.
  .'. Moses was the daughter of Pharaoh's son.

(6) If Aeschines took part in the public rejoicings over the success of my policy, he is inconsistent in condemning it now; if he did not, he was a traitor then.

(7) It is wrong to stick knives into people.
  .'. Surgeons ought to be punished.

(8) If a thing admits of being taught, there must be both teachers and learners of it.
  .'. If there are neither teachers nor learners of a thing, that thing does not admit of being taught.

(9) It is unnecessary to lend books, if they are common, and wrong to lend them, if they are rare. Therefore books should not be lent from public libraries.

(10) Seeing is believing. .'. What is not seen cannot be believed.

(11) St. Paul was not of Jewish blood, for he was a Roman citizen.

(12) To call you an animal is to speak the truth.
  To call you an ass is to call you an animal.
  .'. To call you an ass is to speak the truth.

(13) Pain chastens folly. A life of ease must therefore be one of folly incurable.

(14) We cannot be happy in this world; for we must either indulge our passions or combat them.

(15) It must be clear to the most unlettered mind that, as all things were originally created by the Deity, including the hair on our heads and the beards on our faces, there can be no such thing as property.

(16) The crime was committed by the criminal.
  The criminal was committed by the magistrate.
  .'. The crime was committed by the magistrate.

(17) General councils are as likely to err as the fallible men of whom they consist.

(18) Dead dogs are heavier than living ones, because vitality is buoyant.

(19) Deliberation is concerned with actions.
  Actions are means.
  .'. Deliberation is concerned with means.

(20) 'No beast so fierce but has a touch of pity; But I have none: therefore I am no beast.'

(21) Practical pursuits are better than theoretical.
  .'. Mathematics are better than logic.

(22) Death must be a good. For either the soul, ceasing to be, ceases ta suffer, or, continuing to be, lives in a better state.

(23) What is right should be enforced by law.
  .'. Charity should be so enforced.

(24) All animals were in the Ark.

∴. No animals perished in the Flood.

(25) If he robs, he is not honourable.

If he pays all his dues, he does not rob.

∴. If he pays all his dues, he is honourable.

(26) A dove can fly a mile in a minute.

A swallow can fly faster than a dove.

∴. A swallow can fly more than a mile in a minute.

(27) 'I must soap myself, because it's Sunday.'

'Then do you only soap yourself on Sunday.'

(28) If the charge is false, the author of it is either ignorant or malicious. But the charge is true. Therefore he is neither.

(29) All the angles of a triangle are equal to two right angles.

The angle at the vertex is an angle of a triangle.

∴. It is equal to two right angles.

(30) Si gravis sit dolor, brevis est; si longus, levis. Ergo fortiter ferendus.

(31) You are not what I am.

I am a man.

∴. You are not a man.

(32) The extension of the franchise is necessary, for it is imperative that the right of voting should be granted to classes who have hitherto not possessed this privilege.

(33) If Hannibal is really victorious, he does not need supplies; while, if he is deluding us, we ought certainly not to encourage him by sending them. Livy, xxiii. 13. § 5.

(34) Laws must punish, and punishment hurts.
All laws therefore are hurtful.

(35) The sun is an insensible thing.

The Persians worship the sun.

∴. The Persians worship an insensible thing.

(36) Some ores are not metals; for they are not fluids, and some metals are not fluids.

(37) All the Grecian soldiers put the Persians to flight.

∴. Every Grecian soldier could rout the Persians.

(38) The resurrection of Jesus Christ is either an isolated fact or else admits of parallel. But if it be an isolated fact, it cannot be rendered probable to one who denies the authority of Christianity; and, if it admit of parallel, it no longer proves what is required. Therefore it is either incapable of being substantiated or else makes nothing for the truth of Christianity.

(39) The resurrection of Christ in the flesh and his ascension into heaven were events either intrinsically incredible in their nature or not. If the former, the prevalent belief in them can only be accounted for by miracles; if the latter, they ought to be believed even without miracles. St. Aug. De Civ. Dei, xxii. 8.

(40) Only contented people are wise. Therefore the tramp contented in his rags is necessarily a wise man.

(41) Four-legged things are brutes.

Tables are four-legged things.

∴. Tables are brutes.

(42) The apparent volcanoes in the moon are not volcanoes; for eruptions are produced by gases only, and there are no gases in the moon.

(43) To read the Scriptures is our duty. Therefore the Captain was wrong in punishing the helmsman for reading the Bible at the time when the ship struck.

(44) The divine law orders that kings should be honoured.

Louis Quatorze is a king.

∴. The divine law orders that Louis Quatorze should be honoured.

(45) Those who desire the same object are unanimous.

Caesar and Pompey both desire the same object, namely, supreme power.

∴. They are unanimous.

(46) Either the ministers left at home will be ciphers or they will not be ciphers. If they are ciphers, cabinet government, which is equivalent to constitutional government, will receive a rude blow. If they are not ciphers, the cabinet will be considering matters of the utmost importance in the absence, and the gratuitous absence, of two of its most important members. 'The Standard,' Wed. June 5, 1878.

(47) One patent stove saves half the ordinary amount of fuel. Therefore two would save it all.

(48) One number must win in the lottery.

My ticket is one number.

∴. It must win.

(49) All good shepherds are prepared to lay down their lives for the sheep.

Few in this age are so prepared.

∴. Few in this age are good shepherds.

(50) You cannot define the sun; for a definition must be clearer than the thing defined, and nothing can be clearer than the source of all light.

(51) To give the monopoly of the home market to the produce of domestic industry … must in almost all cases be either a useless or a hurtful regulation. If the produce of domestic can be brought there as cheap as that of foreign industry, the regulation is evidently useless; if it cannot, it is generally hurtful. Adam Smith, Wealth of Nations, Bk. iv. ch. 2.

(52) Verberare est actio. Ergo et vapulare.

(53) The ages of all the members of this family are over 150. The baby is a member of this family. ∴. Its age is over 150.

(54) Romulus must be an historical person; because it is not at all likely that the Romans, whose memory was only burdened with seven kings, should have forgotten the most famous of them, namely, the first.

(55) All scientific treatises that are clear and true deserve attention.

Few scientific treatises are clear and true.

∴. Few scientific treatises deserve attention.

(56) The Conservative Government is an expensive one; for, on their going out of Office, there was a deficit.

(57) A man is forbidden to marry his brother's wife, or, in other words, a woman is forbidden to marry her husband's brother, that is, a woman is directly forbidden to marry two brothers. Therefore a man may not marry two sisters, so that a man may not marry his wife's sister.

INDEX.

The references refer to the sections.

THE END.